SpringerBriefs in Physics

SpringerBriefs in Physics are a series of slim high-quality publications encompassing the entire spectrum of physics. Manuscripts for SpringerBriefs in Physics will be evaluated by Springer and by members of the Editorial Board. Proposals and other communication should be sent to your Publishing Editors at Springer.

Featuring compact volumes of 50 to 125 pages (approximately 20,000–45,000 words), Briefs are shorter than a conventional book but longer than a journal article. Thus, Briefs serve as timely, concise tools for students, researchers, and professionals.

Typical texts for publication might include:

- A snapshot review of the current state of a hot or emerging field
- A concise introduction to core concepts that students must understand in order to make independent contributions
- An extended research report giving more details and discussion than is possible in a conventional journal article
- A manual describing underlying principles and best practices for an experimental technique
- An essay exploring new ideas within physics, related philosophical issues, or broader topics such as science and society

Briefs allow authors to present their ideas and readers to absorb them with minimal time investment.

Briefs will be published as part of Springer's eBook collection, with millions of users worldwide. In addition, they will be available, just like other books, for individual print and electronic purchase.

Briefs are characterized by fast, global electronic dissemination, straightforward publishing agreements, easy-to-use manuscript preparation and formatting guidelines, and expedited production schedules. We aim for publication 8–12 weeks after acceptance.

More information about this series at http://www.springer.com/series/8902

Kim Sharp

Entropy and the Tao of Counting

A Brief Introduction to Statistical Mechanics and the Second Law of Thermodynamics

 Springer

Kim Sharp
Biochemistry and Biophysics
University of Pennsylvania
Philadelphia, PA, USA

ISSN 2191-5423   ISSN 2191-5431   (electronic)
SpringerBriefs in Physics
ISBN 978-3-030-35459-6   ISBN 978-3-030-35457-2   (eBook)
https://doi.org/10.1007/978-3-030-35457-2

This Springer imprint is published by the registered company Springer Nature Switzerland AG.
The registered company address is: Gewerbestrasse 11, 6330 Cham, Switzerland

For Thi and Thuan

Preface

The second law of thermodynamics, which says that a certain quantity called entropy always increases, is the law that scientists believe gives the direction to time's arrow. It is the reason why so many everyday events happen only one way. Heat flows from hot to cold. Stirring milk into coffee always mixes it, never unmixes it. Friction slows things down, never speeds them up. Even the evolution of the universe and its ultimate fate depend on entropy. It is no surprise that there are numerous books and articles on the subjects of entropy and the second law of thermodynamics. So why write another book, even a short one, on the subject?

In 1854 Rudolf Clausius showed that there was a well-defined thermodynamic quantity for which, in 1865, he coined the name entropy. In his words *"The energy of the universe remains constant. The entropy of the universe tends to a maximum."* Entropy could be measured and its behavior described, but what it *was* remained a mystery. Twelve years later Ludwig Boltzmann provided the answer in atomistic, probabilistic terms and helped create the field of statistical mechanics.

Boltzmann's explanation was the first, and arguably still the best, explanation of entropy. His classic paper, published in 1877, is probably never read anymore yet something can still be learned by reading it. One of the things Boltzmann explains very clearly is that there are two equally important contributions to entropy: one from the distribution of atoms in space and the other from the motion of atoms (heat); in other words from the distribution of kinetic energy among the atoms.[1] However, many introductory books and articles on entropy do not give equal attention to each. The spatial contribution by its nature is just easier to visualize and explain. Explanations of the kinetic energy contribution to entropy often use fossilized concepts like Carnot cycles and efficiencies of heat engines. It is also customary to teach thermodynamics before statistical mechanics, because the former is thought to be easier, and this follows their historical order of

---

[1]We now know of four sources of entropy in the universe. The third is radiation: photons have entropy, a fact also discovered by Boltzmann. The fourth is black holes. These require a level of physics and mathematics beyond the scope of this book.

development. Nevertheless, no purely thermodynamic description can explain what entropy actually is. Nor can it explain the origin of the central concept, the famous Boltzmann distribution. This distribution is either introduced by fiat or derived using some fairly advanced mathematical techniques in a later statistical mechanics course which many students do not take. Either way, the result is an incomplete understanding of entropy.

My reason for writing this book is to go back to the beginning, as it were, and give a concise but accurate explanation of entropy and the second law of thermodynamics. I take the contrary position that this can be done better by teaching statistical mechanics *before* (or even without) thermodynamics and that this can be done using just the basic concepts of mechanics: position, velocity, force, and kinetic energy. To do this I resurrect a very useful concept of Boltzmann's, complexion counting. A complexion is simply the list of positions and velocities of all the atoms. I use geometric arguments as much as possible, because these are more conducive to physical understanding than calculus. I also give, for several of the most important entropic effects, estimates of the huge numbers involved, again because this leads to a better physical understanding of entropy. The mathematics requires only a basic understanding of algebra, the exponential function and the logarithm function, at the level of an undergraduate majoring in the physical or biological sciences, or even a high school student with senior level physics.

Finally I want to thank my colleague Franz Matschinsky. His help, wisdom, and good humor were essential in turning my rash decision to translate Boltzmann's seminal 1877 paper on entropy into reality. As a result I was able to read it and fully appreciate the magnitude of Boltzmann's achievements, which in turn led me to write this book.

Philadelphia, PA, USA   Kim Sharp
August 2019

Contents

Chapter 1
Learning to Count

Abstract In 1877 Boltzmann explained the second law of thermodynamics in these terms:

> ... the system will always rapidly approach a more probable state until it finally reaches the most probable state, i.e. that of the heat equilibrium. ... we will be able to identify that quantity which is usually called entropy with the probability of the particular state.

Unpacking the meaning of this quote provides the key to understanding entropy. The term more probable simply means that some states or transformations between states can happen in more ways than others. What we then observe in the real world is that which can happen the most ways. This chapter first defines what is meant by the term way, and then shows how to count these ways. In Chinese philosophy the term Tao can be translated as 'The Way', so we may say that subject of this chapter is the Tao of counting.

Keywords Velocity · Kinetic energy · Temperature · Complexions · Friction · Partition function

1.1 A Simple Analogy

As an introduction to this business of counting, consider rolling two dice. A possible outcome is a roll of 2 with dice number one, and a roll of 1 with dice number two, which will be abbreviated as $(2, 1)$ (Fig. 1.1). Another outcome is a roll of $(1, 2)$. This is a different outcome from the first roll because which dice has which roll is different. Each possible outcome, defined as the specific values for each dice, will be called a complexion. There are $6 \times 6 = 36$ possible complexions. Now we apply a condition, for example that the spots on the dice must sum to a fixed value, and ask how many complexions satisfy that condition. If we require the sum to equal 7 for example, there are six complexions that satisfy this condition: $(1, 6)$, $(2, 5)$, $(3, 4)$, $(4, 3)$, $(5, 2)$, $(6, 1)$. In contrast, a total of say 12 can be realized in just one way,

Fig. 1.1 Two dice, two
complexions

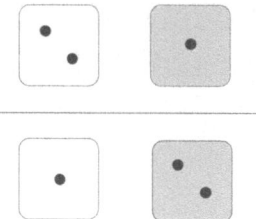

(6, 6). The lesson from the dice analogy is that different conditions can be realized
by different numbers of complexions. We are very close to the meaning of entropy
when we recognize that in a large number of rolls of two dice we are likely to see
a total of 7 more often than any other total simply because it is produced by more
complexions than any other total. To understand thermodynamics, we need only to
apply the same ideas to atoms.

1.2   Atoms Have Position and Velocity

Since we live in a three-dimensional universe we need three coordinates, x, y
and z, to describe the position of an atom, as shown in the left panel of Fig. 1.2.
In addition, all atoms are in constant motion because they have thermal or heat
energy. More precisely, they have kinetic energy. Even atoms in a solid are moving
rapidly back and forth. To describe the velocity of an object, we must not only
say how fast it is moving—its speed in meters per second—but in what direction.
For example, an airplane might be traveling west at 100 m/s, climbing at 10 m/s,
and drifting north at 1 m/s because of a cross wind. To completely describe the
motion of an atom we need to specify its speed in each of the x, y and z directions.
These are the velocity components, u, γ and w illustrated in the right panel of
Fig. 1.2.

The magnitude of the atom's velocity, or its speed, can be calculated using
Pythagoras' theorem as

$$v = \sqrt{u^2 + \gamma^2 + w^2}. \tag{1.1}$$

v is the length of the velocity arrow in Fig. 1.2. In summary, at any moment in time
an atom is described by three numbers for its position, and three numbers for its
velocity, which will be abbreviated as a list of six numbers (x, y, z, u, γ, w).

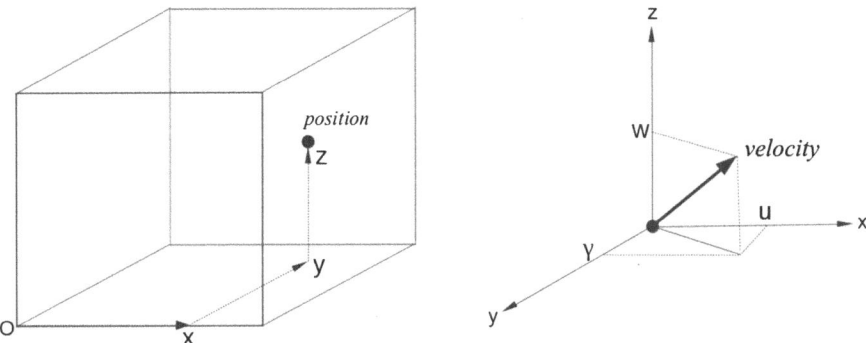

Fig. 1.2 Describing an atom by position and velocity

1.3 The Kinetic Energy of an Atom

An atom that is in motion has a kinetic energy proportional to its mass m, and proportional to the square of its speed v

$$E_k = \frac{1}{2}mv^2. \tag{1.2}$$

Because the kinetic energy of an atom depends on the square of its speed the total kinetic energy is the sum of the contributions from the x, y and z components of velocity:

$$E_k = \frac{1}{2}mu^2 + \frac{1}{2}m\gamma^2 + \frac{1}{2}mw^2 \tag{1.3}$$

This can be verified by substituting for v in Eq. 1.2 using Eq. 1.1. So we can speak of an atom as having three independent kinetic energy components.

If we have N atoms, the total kinetic energy is the sum of the atomic kinetic energies

$$E_K = \frac{1}{2}\sum_i^N m_i(u_i^2 + \gamma_i^2 + w_i^2) \tag{1.4}$$

where the subscript i labels the atom. The mean kinetic energy per atom is $\hat{E}_K = E_K/N$. If N is large then the mean kinetic energy per atom and the absolute temperature, T, are just the same thing but expressed in different units. In fact Maxwell and Boltzmann, when developing statistical thermodynamics, mostly used mean energy per atom not temperature.[1] Because, in the development of thermodynamics, the units of energy and the units of absolute temperature were determined independently the scale factor relating them had to be determined empirically. This was done, for example, using the properties of an ideal gas and Boyle's Law (see Appendix). It required knowing how many atoms are in a given amount of material, i.e. knowing Avogadro's constant. The relationship is $k_b T = \frac{2}{3}\hat{E}_K$, where k_b is called Boltzmann's constant. If energy is measured in Joules and temperature is in degrees Kelvin then $k_b = 1.38 \times 10^{-23}$ J/K.

1.4 A Complexion of Atoms

If we have a collection of N atoms, at any moment in time we need $3N$ coordinates (x_i, y_i, z_i) to specify the position of all the atoms in space and $3N$ velocity components (u_i, γ_i, w_i) in the x, y and z directions, respectively, to specify the motion of all the atoms. The set of all $6N$ position-velocity coordinate values,

$$(x_1, y_1, z_1, u_1, \gamma_1, w_1 \ldots x_N, y_N, z_N, u_N, \gamma_N, w_N) \tag{1.5}$$

defines what will be called a *complexion*. The term complexion is the Anglicized form of the word Komplexion used by Boltzmann in his original work explaining entropy (Boltzmann 1877; Sharp and Matschinsky 2015). I resurrect this neutral term to avoid baggage-laden terms such as microstate, disordered state, etc. often used in explaining entropy. Because the atoms are in motion the system will be continually moving from one complexion to another according to the laws of dynamics. The set of $3N$ position coordinate values alone, $(x_1, \ldots z_N)$, will be referred to as a spatial complexion. Similarly, the set of $3N$ velocity coordinate values alone, $(u_1, \ldots w_N)$, will be referred to as a velocity complexion, or equivalently, a kinetic energy complexion, since the kinetic energy components of the atoms are completely specified by their velocity components. Figure 1.3 shows an example of four of the many possible complexions that three atoms could adopt.

---

[1]In their papers Maxwell and Boltzmann used the symbol T to refer to kinetic energy, not temperature; a convention physicists still use today and a source of confusion.

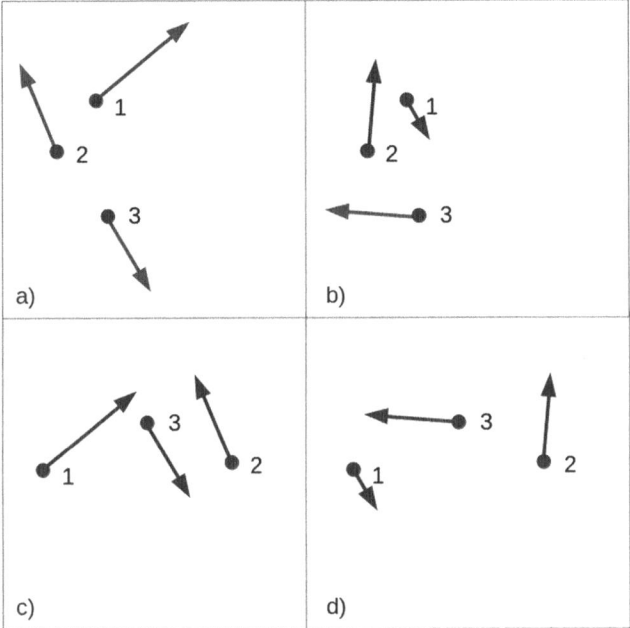

Fig. 1.3 Four possible complexions of three atoms. (**a**) and (**b**) have the same spatial complexion, but different velocity complexions, as do (**c**) and (**d**). (**a**) and (**c**) have the same velocity complexion, but different spatial complexions, as do (**b**) and (**d**). (**a**) and (**d**) have different spatial and velocity complexions, as do (**b**) and (**c**)

1.5 Boltzmann's Postulate

Now that a complexion has been defined we introduce Boltzmann's fundamental postulate.

> Every complexion of N atoms having the same total energy E is equally likely

This is a postulate; it is not derived from any more fundamental principle. It is, however, motivated by the principle of insufficient reason in the absence of any definite knowledge to prefer one complexion over another (Bridgman 1972). Its ultimate justification must be how well the resulting theory explains observed thermodynamic behavior (Tolman 1938). Since statistical mechanics, within its domain of applicability, has proven to be an extremely reliable theory we have no reason to doubt Boltzmann's fundamental postulate.

The next question, then, is how many complexions are there for a given number of atoms with a given total energy? How can we count the 'ways' of arranging the atoms and their velocities? This task is described in the next section. With Boltzmann's postulate and the ability to count complexions we will then derive the Second Law of Thermodynamics and the properties of entropy.

1.6   Counting Complexions: Part 1

The laws of dynamics describe how, given a set of atomic positions and velocities at some instant, these positions and velocities change with time. These laws do not, however, specify what positions and velocities are possible, only how they change. Without violating the laws of dynamics, any given spatial arrangement of atoms can in principle be combined with any given set of atomic velocities. In the language of statistical mechanics positions and velocities are independent degrees of freedom (see, for example, Fig. 1.3). So we can split the task of complexion counting into two parts. First, that of counting all possible velocity complexions with the same total kinetic energy E_K for any given spatial complexion. Second, accounting for the different spatial complexions including, if necessary, any change in potential energy with atomic positions. In this section the machinery for counting velocity complexions is derived using a geometric argument and basic concepts of mechanics.

1.6.1   A Single Atom

Consider an atom i with velocity components (u, γ, w). It has kinetic energy

$$E_{k,i} = \frac{1}{2} m_i (u^2 + \gamma^2 + w^2). \tag{1.6}$$

The velocity of this atom can represented by a vector arrow which has a length, using Pythagoras' theorem, of $v_i = \sqrt{2E_{k,i}/m_i}$ (Fig. 1.4). There are many other combinations of x, y and z component velocities, (u', γ', w'), that have the same kinetic energy provided that they satisfy the equation $E_{k,i} = \frac{1}{2} m_i (u'^2 + \gamma'^2 + w'^2)$. Each possibility corresponds to a velocity vector of the same length (magnitude) v_i but representing a different velocity complexion. Every possible velocity complexion with the same kinetic energy can be represented by a vector starting at the origin and ending somewhere on the surface of a sphere of radius $R = \sqrt{2E_{k,i}/m_i}$. Figure 1.4 shows two other velocity vectors of the same magnitude whose velocities differ by a small amount δv. It is clear from construction of this 'velocity sphere' that the number of different velocity complexions W_v with fixed kinetic energy $E_{k,i}$ for atom i is proportional to the surface area of its velocity-sphere:

$$W_v \propto R^2 \propto (\sqrt{E_{k,i}})^2. \tag{1.7}$$

1.6.2 Many Atoms

Now consider N atoms with some velocity complexion $(u_1, \gamma_1, \ldots \gamma_N, w_N)$. Initially we will consider the special case where all the atoms have the same mass m, and then generalize to atoms of different masses. The total kinetic energy is

$$E_K = \frac{1}{2}m \sum_{i=1}^{N}(u_i^2 + \gamma_i^2 + w_i^2). \tag{1.8}$$

We can represent this velocity complexion as a vector in 3N-dimensional space with components $(u_1, \gamma_1, \ldots w_N)$. By the 3N-dimensional version of Pythagoras' theorem this vector has length $\sqrt{2E_K/m}$. This is just the generalization of the three-dimensional geometric picture of Fig. 1.4 to 3N dimensions. Then the number of possible velocity complexions, W_v, with the *same* total kinetic energy is proportional to the surface area of a 3N-dimensional velocity-sphere of radius $R = \sqrt{2E_K/m}$. The surface area of a 3N-dimensional sphere, A_{3N}, is proportional to R^{3N-1}. In everything that follows we will be assuming that there is a large number of atoms, so we will now make the approximation $3N - 1 \approx 3N$ to make the equations tidier and the math easier to follow. So $A_{3N} \propto R^{3N-1} \approx R^{3N}$. Since $R \propto E_K^{1/2}$ the number of possible velocity complexions scales as

$$W_v \propto E_K^{3N/2} \tag{1.9}$$

with respect to the total kinetic energy. Now obviously the absolute number of complexions depends on how finely we divide up the surface of the 3N-dimensional sphere; how small is the surface area element δA_{3N} which represents

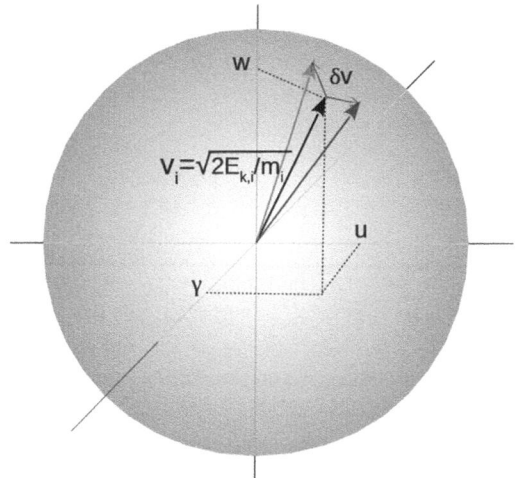

Fig. 1.4 Possible velocities of a single atom

a single velocity complexion. However, all we will need are ratios of numbers of complexions corresponding to some change in E_K. So the size of the area element δA_{3N} will cancel out. To calculate changes in the number of complexions the only thing that matters is that this number scales as $E_K^{3N/2}$.

1.6.3 The Effect of Heat on Velocity Complexions

One of Boltzmann's most important discoveries was how to compute the change in number of complexions when heat is added to the system (Boltzmann 1877). His original derivation, which is followed by most textbooks of statistical mechanics, is quite involved mathematically and is thus hard to follow for the non-specialist. So here we pursue the geometric argument of Fig. 1.4. If a small amount of heat δQ is added to a collection of N atoms, the kinetic energy increases from E_K to $E_K + \delta Q$. The radius of the $3N$-dimensional sphere representing all possible velocity vectors increases by a factor of

$$\frac{R'}{R} = \left(\frac{E_K + \delta Q}{E_K}\right)^{1/2} = \left(1 + \frac{\delta Q}{N\hat{E}_K}\right)^{1/2} \tag{1.10}$$

where $\hat{E}_K$ is the mean kinetic energy per atom. The surface area of this sphere increases by a factor of

$$\left(\frac{R'}{R}\right)^{3N} = \left(1 + \frac{\delta Q}{N\hat{E}_K}\right)^{3N/2}. \tag{1.11}$$

The number of complexions increases by the same factor as the surface area:

$$\frac{W'_v}{W_v} = \left(1 + \frac{\delta Q}{N\hat{E}_K}\right)^{3N/2}, \tag{1.12}$$

which can be re-written as

$$\frac{W'_v}{W_v} = exp\left(\frac{3N}{2}ln(1 + \delta Q/N\hat{E}_K)\right) \tag{1.13}$$

using the identity relationship between exponentials and logarithms: $x = exp(lnx)$, and the fact that $ln(x^n) = nln(x)$. See the Appendix.

Since N is large, $\delta Q << N\hat{E}_K$ and using the fact that $ln(1 + x) \approx x$ when x is small (Appendix), we can make the approximation:

$$ln(1 + \delta Q/N\hat{E}_K) \approx \delta Q/N\hat{E}_K. \tag{1.14}$$

Substituting this approximation into the exponent gives Boltzmann's equation

$$\frac{W'_v}{W_v} = exp\left(\frac{3}{2}\frac{\delta Q}{\hat{E}_K}\right) = exp\left(\frac{\delta Q}{k_b T}\right) \tag{1.15}$$

where the second equality uses the relationship of Sect. 1.3 between temperature and mean kinetic energy, namely $k_b T = \frac{2}{3}\hat{E}_K$.

1.6.4 Accounting for Atoms with Different Masses

We now generalize the argument leading to Eq. 1.15 for atoms with different masses. Equation 1.8 for the total kinetic energy is now replaced by

$$E_K = \frac{1}{2}\sum_{i=1}^{N} m_i(u_i^2 + \gamma_i^2 + w_i^2). \tag{1.16}$$

Equation 1.16 can be rewritten as

$$1 = \sum_{i=1}^{N}(\frac{u_i^2}{a_i^2} + \frac{\gamma_i^2}{a_i^2} + \frac{w_i^2}{a_i^2}) \tag{1.17}$$

where $a_i = \sqrt{2E_K/m_i}$. This is the equation for a 3N-dimensional ellipsoid with semi-axes of length a_i. For the same kinetic energy a lighter atom has a greater velocity than a heavy atom. So the 3N-dimensional velocity-sphere becomes elongated along the axial directions of lighter atoms, and shortened along the directions for heavier atoms, becoming an ellipsoid. All possible velocity complexions for fixed E_K are represented by a 3N-dimensional vector from the center of this ellipsoid ending somewhere on its surface. The argument about how the number of complexions depends on E_K is the same as for the equal mass/sphere case: W_v is given by how many different 3N-dimensional velocity vectors there are, which is proportional to the surface area of the 3N-ellipsoid. This ellipsoid is of fixed shape determined by the various masses of the atoms. Thus its surface area scales as the $(3N-1)^{th} \approx (3N)^{th}$ power of the linear dimension of the ellipsoid. The linear dimension scales as $\sqrt{E_K}$, so the area scales as the $\frac{3N}{2}^{th}$ power of E_K as before. The addition of a small amount of heat provides the same factor of $(1 + \delta Q/N\hat{E}_K)^{3N/2}$ for the number of complexions. For $\delta Q << N\hat{E}_K$ this leads again to Boltzmann's equation 1.15.

1.6.5 The Meaning of Boltzmann's Equation

Equation 1.15 says that if a small amount of heat (kinetic energy) δQ is added to N atoms that have average kinetic energy $\hat{E}_K$, then the number of velocity complexions increases by a factor of W_v'/W_v given by Eq. 1.15 where W_v and W_v' are the number of velocity complexions before and after the heat addition. The increase in number of velocity complexions comes from the increase in the number of ways to distribute the kinetic energy among the N atoms. Conversely, if we remove heat (δQ is negative) the number of complexions decreases. The exponential form of Eq. 1.15 is a reflection of the combinatoric nature of complexion counting. For example, if we add an amount of heat equal to the average kinetic energy of one atom, $\hat{E}_K$, the increase in velocity complexions is a factor of $e^{3/2} = 4.48$. If twice as much heat is added, then the number of complexions increases by the *square* of this factor, about 20-fold.

Another feature of Boltzmann's equation is that the effect of a given increment of heat depends on how much heat is already there; how large δQ is compared to the average kinetic energy of the atoms. In terms of the geometric picture of Fig. 1.4, the increment in radius of the velocity sphere due to a fixed amount of heat δQ increases the surface area of the 3N-sphere by a larger factor when the radius of the sphere (which is proportional to $\sqrt{E_K}$) is smaller. The proportionally larger effect of heat added at a lower temperature is graphically illustrated in Fig. 1.5 using the velocity sphere of a single atom. Here 2 units of kinetic energy are added to an atom starting with 4 units of kinetic energy (low T) or starting with 8 units of energy (high T). It is apparent from the figure, which is drawn to scale, that the area of the low T sphere (on the left) is increased by a larger factor than the high T sphere.

Most of what follows will make use of Boltzmann's equation 1.15. Before we move on to complete the counting of complexions, the next section gives two applications of this equation to demonstrate its explanatory power.

Fig. 1.5 Effect of adding heat is greater at lower temperature

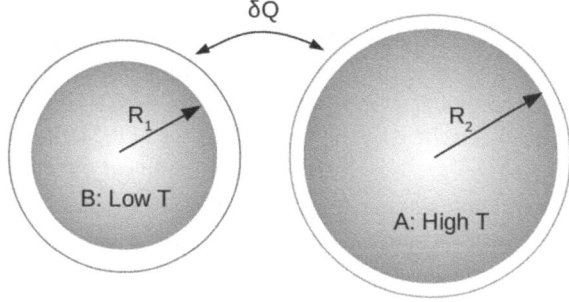

1.7 Why Heat Flows from Hot to Cold

If we examine Boltzmann's equation 1.15 we see that in the exponent the amount of heat δQ is divided by the mean kinetic energy per atom, $\hat{E}_K$. A given amount of heat will cause a greater increase in the number of velocity complexions (ways to distribute the kinetic energy) when it is added to a collection of atoms with a smaller amount of kinetic energy, i.e. with a lower temperature (see Fig. 1.5). So if a small amount of heat leaves a body A at temperature T_A and enters a body B at a lower temperature T_B it is clear that the number of velocity complexions for A decreases by a smaller factor than that by which the number of velocity complexions for B increases, because $T_B < T_A$. The total number of velocity complexions is the product of the complexion numbers of A and B, so the total changes by a factor

$$\frac{W'}{W} = \frac{W'_A W'_B}{W_A W_B} = exp\left(\frac{\delta Q}{k_b}\left(\frac{1}{T_B} - \frac{1}{T_A}\right)\right). \tag{1.18}$$

This factor is greater than one because $1/T_B > 1/T_A$ and the exponent is positive. If heat flowed in the reverse direction, from cold to hot, the number of complexions would decrease, which by Boltzmann's fundamental postulate means it is less likely to be observed. So heat will flow from A to B until the number of complexions stops increasing, i.e. reaches a maximum. This happens when $T_B = T_A$, which is thermal equilibrium. Put another way, when $T_B = T_A$, the exponent in Eq. 1.18 is zero, so the ratio W'/W is one. The transfer of a small amount of heat in either direction does not change the number of velocity complexions.

The real import of counting complexions comes when representative numbers are calculated. Take two 0.1 kg blocks of copper. One is at a temperature of 300 K. The other is *one millionth of a degree* hotter at 300.000001 K. From the heat capacity of copper, 385 J/kg/K, we can calculate how much more heat the hotter block has: 38.5 μJ. To equalize the temperature, half of this extra heat must flow into the colder block. Evaluating Eq. 1.18 with $\delta Q = 19.25\,\mu$J and $k_b = 1.38 \times 10^{-23}$ J/K the exponent has a value of about 7.8×10^6, giving

$$\frac{W'}{W} = e^{7.8 \times 10^6} \approx 10^{3,400,000}. \tag{1.19}$$

The number of kinetic energy complexions increases by a factor of 1 followed by more than *three million zeros*.[2] The equal temperature situation has overwhelmingly more complexions. Clearly the possibility of anything more than a really microscopic flow of heat in the reverse direction, from cold to hot, is close to zero.

---

[2]Using $e^x = 10^{(x/ln10)}$, $ln\,10 = 2.303$. See Appendix.

[*Technical note:* Equation 1.15 was derived assuming that the change in temperature upon addition of the heat was negligible, while of course in the above thermal equilibration example the temperature of A does change, from T_A to $T_A - \Delta T/2$, while that of B changes from T_B to $T_B + \Delta T/2$, where $\Delta T = T_A - T_B$. So the calculation was done assuming constant temperatures for A and B equal to their "average values" during the equilibration, namely $T_A - \Delta T/4$ and $T_B + \Delta T/4$. For this example the error compared to the exact result (obtained by integration over a continuous temperature change) is tiny.]

1.8 Why Friction Slows Things Down

A rubber ball is released from a certain height above a smooth hard surface. It accelerates downwards, Fig. 1.6, hits the surface, bounces up a few times, each time to a lesser height until finally it comes to rest on the surface. Since energy is conserved why does this happen, why does the ball not bounce back to the original height and keep on bouncing for ever? Or why cannot the ball, after remaining at rest for a while, spontaneously move upwards to its initial height? Initially the ball has a certain potential energy mgh where m is the mass of the ball, h is the release height above the surface, and g is the acceleration due to Earth's gravity. This potential energy is converted into the kinetic energy of the ball's motion, $E_K = \frac{1}{2}mv^2 = mgh$, where v is the speed of the ball when it reaches the surface. How many ways are there to distribute this kinetic energy among the atoms of the ball? In effect, just one, where all atoms have identical velocity components $(u, \gamma, w) = (0, 0, -v)$ equal to the center of mass motion. When the ball comes to rest this kinetic energy has been converted to exactly the same amount of heat: kinetic energy now distributed at random among the atoms of the ball, the surface, the air.

Boltzmann's equation 1.15 tells us how the number of velocity complexions has increased. For example take a 0.1 kg ball released from a height of 0.1 m, at room temperature, 300 K. The gravitational constant is $g = 10\,\text{m/s}^2$, so $\delta Q = 0.1\,\text{J}$. The

Fig. 1.6 Friction slows objects down

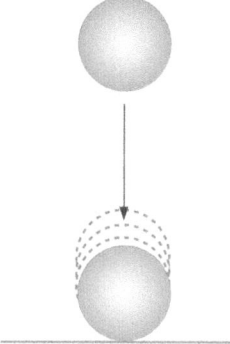

exponent in Eq. 1.15 is 2.42×10^{19}. The number of complexions increases by a factor

$$\frac{W'}{W} = e^{2.42 \times 10^{19}} \approx 10^{10^{19}}. \tag{1.20}$$

In other words, the number of complexions increases by a factor of 1 followed by *ten quintillion* (10^{19}) zeros. This is a mind-bogglingly large number. No wonder the effect of friction is irreversible and the kinetic energy components in the surroundings could never spontaneously 'line up' to push the ball back up in the air once it has come to rest.

1.9 Why Is the Number of Complexions So Large?

In the previous two sections we encountered some extremely large numbers. It is helpful to understand what is the origin of these large numbers, and just how large they are. This is key to understanding why the Second Law of Thermodynamics, even though it is statistical, is called a law.

The large numbers are a consequence of one physical factor and one mathematical factor. On the physical side, atoms are extremely small, which is reflected in the fact that Avogradro's number is large: 6.02×10^{23} molecules per mole. This means that even a nanoscale amount of matter contains a lot of atoms. One billionth of a gram of diamond, for example, still contains about 5×10^{13} atoms of carbon, although this number is unlikely to impress one's fiancée! The mathematical side involves permutations, which describe the number of ways of arranging things (here atoms or atomic velocities). The number of permutations increases very rapidly with the number of things. Roughly speaking permutations scale as n^n, where n is the number of things. We say that a number like e^n or 10^n, where n occurs in the exponent, increases literally exponentially with n. But with permutations n occurs as the base *and* the exponent, so permutations increase supra-exponentially. Put these two factors together and numbers like $10^{3.4 \times 10^6}$ and $10^{10^{19}}$ are the result.

It is not accurate to call these numbers astronomical, because astronomical numbers like the number of atoms in the universe (on the order of 10^{79}), are by comparison small. Even a googol, the archetypal large number, has only one hundred zeros after the 1, and so it is also a piker in comparison. A better yardstick is the proverbial monkey plinking away randomly on a typewriter and producing the exact text of Hamlet. Say Hamlet contains about 27,000 characters, including spaces and punctuation. We'll be strict and want correct capitalization and punctuation, so there are 26 upper and lower case characters to choose from, plus the space and say seven punctuation symbols, for a round total of 60 possible symbols at each of 27,000 positions. The odds are $1:60^{27,000} \approx 1:10^{48,000}$ against the monkey producing Hamlet. But this involves a number which is not even close to the smaller of the two complexion numbers occurring in the previous two examples.

1.10   Counting Complexions: Part 2

In Sect. 1.6 we derived the number of velocity complexions as a function of the total kinetic energy for any given spatial complexion. We now add in the spatial complexion contribution to provide a complete prescription for complexion counting.

1.10.1   Combining Spatial and Velocity Complexions

We again have a collection of N atoms with total kinetic energy E_K. First assume that there are no forces acting on the atoms, either from other atoms or from external factors. This means that there is no variation in potential energy with the positions of the atoms. In this case if there are W_s spatial complexions each can be combined with all W_v possible velocity complexions having the same total kinetic energy E_K. This gives a total of

$$W = W_s \times W_v \tag{1.21}$$

complexions. An example in miniature is shown in Fig. 1.3. Here there are two different spatial complexions which are combined with two different velocity complexions to give a total of four complexions. If we apply Boltzmann's postulate to Eq. 1.21, what it tells us is that in the absence of forces each spatial complexion is equally likely, since each can be combined with exactly the same number of velocity complexions.

Now assume that there are forces acting. Equation 1.21 no longer applies. When an atom moves its potential energy is changed according to the relationship

$$\delta U = -f\delta x \tag{1.22}$$

where δx is the displacement along the direction of a force of magnitude f. If the atom moves in the direction of the force its potential energy decreases, in other words it moves downhill in potential energy. This energy is converted into kinetic energy as the atom is accelerated according to Newton's Third Law: $f = ma$. The reverse is true if the atom moves against the force. The atom slows down and loses kinetic energy as it moves uphill in potential energy. Since energy is conserved and there are only two types of energy, kinetic and potential, we have

$$\delta Q = -\delta U.$$

It is immaterial whether the forces are external or come from interactions between the atoms: When there is a change in potential energy heat is created or annihilated, depending on the sign of the change. Because of this heat change, what Boltzmann's key discovery, Eq. 1.15, tells us is that the corresponding spatial complexions are now no longer equally likely, because each has a different number of possible

Fig. 1.7 At higher potential energy atoms have fewer velocity complexions

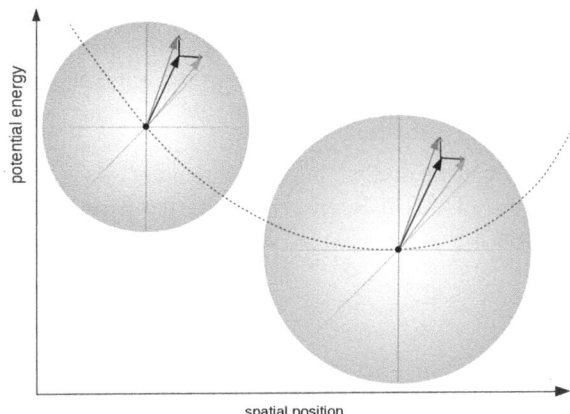

velocity complexions available to it. With constant total energy, spatial complexions of higher potential energy have lower kinetic energy. Therefore they have less kinetic energy complexions and so they are less likely, and vice versa. This is illustrated qualitatively in Fig. 1.7 for the velocity sphere of a single atom at two positions of different potential energy.

To quantify the effect of potential energy on complexions in the general case, first compare two spatial complexions of N atoms, labelled A and B which have different potential energies: U_A and U_B. When the atoms move from complexion A to complexion B an amount of heat $-(U_B - U_A)$ is created if we move downhill in potential energy $(U_B < U_A)$ or is annihilated if we move uphill in potential energy $(U_B > U_A)$. Using Eq. 1.15 the ratio of the corresponding number of velocity complexions is

$$\frac{W_v(B)}{W_v(A)} = exp\left(-\frac{(U_B - U_A)}{k_b T}\right) \qquad (1.23)$$

Thus the two spatial complexions A and B actually correspond to $W(A) = 1 \times W_v(A)$ and $W(B) = 1 \times W_v(B)$ total complexions, respectively. By Boltzmann's fundamental postulate, the relative probability of being in the spatial complexion B vs. the spatial complexion A is given by the ratio of their total numbers of complexions

$$\frac{p_B}{p_A} = exp\left(-\frac{(U_B - U_A)}{k_b T}\right) \qquad (1.24)$$

Put another way, the presence of potential energy has introduced a coupling between the spatial and velocity complexions. From Eq. 1.24 it follows that we can write the absolute probability of any spatial complexion i as

$$p_i = \frac{exp(-U_i/k_b T)}{\sum_{i=1}^{W_s} exp(-U_i/k_b T)} \qquad (1.25)$$

since any pair of probabilities p_i and p_j given by Eq. 1.25 satisfy Eq. 1.24, and the probabilities sum to one. The probability distribution p_i is called the Boltzmann distribution. The term e^{-U_i/k_bT} in the numerator is called the Boltzmann factor. The sum of these factors in the denominator is taken over all possible spatial complexions W_s. This sum functions as a probability normalizing factor and in spite of its mundane origin it plays an important role in entropy and statistical mechanics. Because of this it is given a name, the partition function, often denoted by the symbol Z from the German Zustandssumme (literally sum of states):

$$Z = \sum_{i=1}^{W_s} exp(-U_i/k_bT) \tag{1.26}$$

Using Eq. 1.25 we can now calculate the average value of any property X using

$$X = \sum_{i=1}^{W_s} p_i X_i = \frac{1}{Z} \sum_{i=1}^{W_s} X_i \, exp(-U_i/k_bT) \tag{1.27}$$

where X_i is the value of the property in the ith spatial complexion. For example the mean potential energy is given by

$$U = \sum_{i=1}^{W_s} p_i U_i = \frac{1}{Z} \sum_{i=1}^{W_s} U_i \, exp(-U_i/k_bT) \tag{1.28}$$

1.10.2   The Total Number of Complexions

To obtain the total number of complexions W available to a fixed number of atoms N with fixed total energy E we need to sum the number of velocity complexions available to each spatial complexion:

$$W = \sum_{i=1}^{W_s} W_v(i) \tag{1.29}$$

where in general $W_v(i)$ will vary from spatial complexion to spatial complexion depending on that complexion's potential energy. We will pick an arbitrary spatial complexion o, with potential energy U_o as a reference and use Eq. 1.23 to write the number of velocity complexions possible with spatial complexion i in terms of the number of velocity complexions of the reference state:

$$W_v(i) = W_v(o) \, exp\left(\frac{-(U_i - U_o)}{k_bT}\right) \tag{1.30}$$

The total number of complexions is then given by

$$W = W_v(o)\, exp(U_o/k_bT) \sum_{i=1}^{W_s} exp(-U_i/k_bT) = C'Z \qquad (1.31)$$

The factor coming from the arbitrary complexion o used as a reference state is common to all the spatial complexions. It is brought outside the sum as a constant, C'. The sum itself is just the partition function Z of Eq. 1.26.

1.10.3 Constant Temperature Conditions

Equation 1.31 gives the total number of complexions for a fixed number of atoms with a constant total energy E. This assumes that the collection of N atoms we are interested in is completely isolated from its environment: No heat can enter or leave, which is known as adiabatic conditions. The more common situation is a system in thermal equilibrium with its surroundings, i.e. under constant temperature conditions. How does this affect the counting of complexions? We can apply the results from the heat flow example, Sect. 1.7 to find out. Consider our collection of N atoms at equilibrium with its surroundings at a temperature T. At a certain moment this collection has total energy E. The total number of complexions is given by Eq. 1.31. Since it is in thermal equilibrium with its surroundings, say a small fluctuation (due to collisions with its surroundings, or thermal radiation emitted or received) causes a small amount of heat δQ to pass from the surrounding environment to the system or vice versa. Although Eq. 1.15 says that the number of complexions for the system of N atoms changes, application of Eq. 1.18 with $T_A = T_B = T$ shows that the number of complexions of the surroundings changes by an equal and opposite amount, so that the total number of complexions (of system plus environment) is unaffected by the heat flow allowed by constant T conditions: Complexion counting is the same as though we kept the system at constant energy E.

Summary
Complexion counting has produced the following results

- Equation 1.9: The number of kinetic energy complexions is proportional to the total kinetic energy raised to the $3N/2$th power,

$$W_v \propto E_K^{3N/2}$$

where N is the number of atoms.

(continued)

- Equation 1.15: If a small amount of heat δQ is added, the number of kinetic energy complexions is increased by the factor

$$\frac{W'_v}{W_v} = exp\left(\frac{\delta Q}{k_b T}\right)$$

- Equation 1.21: When there are no forces acting the total number of complexions is the product of the number of spatial and velocity complexions

$$W = W_s \times W_v$$

  and all spatial complexions are equally likely.
- Equation 1.25: When there are forces acting there are potential energy differences, and the probability of a spatial complexion i depends on the potential energy of that complexion and the temperature as

$$p_i = \frac{exp(-U_i/k_b T)}{\sum_{i=1}^{W_s} exp(-U_i/k_b T)}$$

- Equation 1.31: The total number of complexions is proportional to the partition function

$$W \propto \sum_{i=1}^{W_s} exp(-U_i/k_b T)$$

  where the sum is over all possible spatial complexions, and the terms in the sum are the Boltzmann factors.

With the ability to count complexions and determine their probabilities we can now explain what we see at equilibrium, by determining what state has the most complexions. To paraphrase Feynman (1972) much of statistical mechanics is either leading up to the summit of Eqs. 1.25 and 1.27 or coming down from them. In this chapter on complexion counting we proceeded entirely without the concept of entropy. Historically, however, entropy was characterized as a thermodynamic quantity before its molecular, statistical origins were known. So in the next chapter we connect the concepts of complexion counting and entropy.

Chapter 2
What Is Entropy?

Abstract In this chapter Clausius' 1865 macroscopic, thermodynamic definition of entropy is united with Boltzmann's 1877 atomistic, probabilistic definition of entropy.

Keywords Equilibrium · Entropy · Ideal gas · Barometric equation · Maxwell-Boltzmann distribution

2.1   The Statistical, Atomistic Definition of Entropy

In principle, Boltzmann's postulate combined with the methods just described for counting complexions are enough to understand equilibrium thermodynamics. Historically, however, Clausius had already defined a quantity he called entropy, and he showed by thermodynamic arguments that the entropy change in any substance produced by a small amount of added heat is given by

$$\delta S = \frac{\delta Q}{T} \tag{2.1}$$

where δQ is small enough that it causes a negligible change in temperature. But this ratio of added heat to temperature is just what appears in the exponent of Eq. 1.15. If we substitute Eq. 2.1 in and take the logarithm of both sides we get:

$$\delta S = k_b ln \left(\frac{W'_v}{W_v} \right) \tag{2.2}$$

The change in entropy is due to the change in the number of complexions; the change in the number of ways to distribute the kinetic energy among the atoms. The equivalence in Eq. 2.2 led Boltzmann to a more general definition of the entropy of a state in terms of the total number of complexions for that state

$$S = k_b ln W + C \tag{2.3}$$

K. Sharp, *Entropy and the Tao of Counting*, SpringerBriefs in Physics,
https://doi.org/10.1007/978-3-030-35457-2_2

where W includes both spatial and velocity complexions. The more general expression for a change in entropy is then

$$\Delta S = k_b ln \left(\frac{W'}{W} \right) \tag{2.4}$$

The use here of the notation ΔS instead of δS indicates that Boltzmann's expression is valid for any size change of entropy.

C in Eq. 2.3 is an undetermined constant which is unimportant since we are always interested in changes in entropy, so it is convenient to set $C = 0$. Then the logarithmic relationship between entropy and complexions means that entropies of independent systems are additive because the total number of complexions is the product of each system's complexion number.

In Eqs. 2.2–2.4 the units of entropy are energy per degree of temperature, e.g. J/mol/K. So the numerical value depends on the units being used. However, S/k_b is dimensionless. Only dimensionless quantities have physical meaning, but more commonly units of energy per degree Kelvin are used. The crucial point is that entropy is simply a logarithmic measure of the number of complexions. As we have seen it is often insightful to express results directly in terms of numbers of complexions rather than in entropy units: Although the molecular definition of entropy is probabilistic, appreciation of the numbers involved more clearly conveys the almost deterministic nature of entropic effects at the observable level.

2.2   Entropy Maximization Explains Equilibrium Behavior

Thermodynamics deals with quantities like pressure, the distribution of gases or liquids and equilibrium constants of chemical reactions. These are average quantities involving large numbers of atoms, what are called macroscopic quantities. In this context Boltzmann's postulate about equi-probable complexions simply means:

> The macroscopic state that we see at equilibrium is the state that can happen in the most ways; that corresponds to the most complexions.

This principle is illustrated in miniature by the dice analogy: If you roll two dice, the 'macroscopic' total of 7 is the most likely because 6 of the 36 complexions, (1,6), (2,5), (3,4), (4,3), (5,2) and (6,1) produce this total, more complexions than for any other total. We will observe a total of 7 on average one sixth of the time.

Equation 2.3 formalizes Boltzmann's identification of the most probable state, the state that can happen the most ways, with the state of maximum entropy. What must be appreciated is that when you deal with the typical number of atoms involved

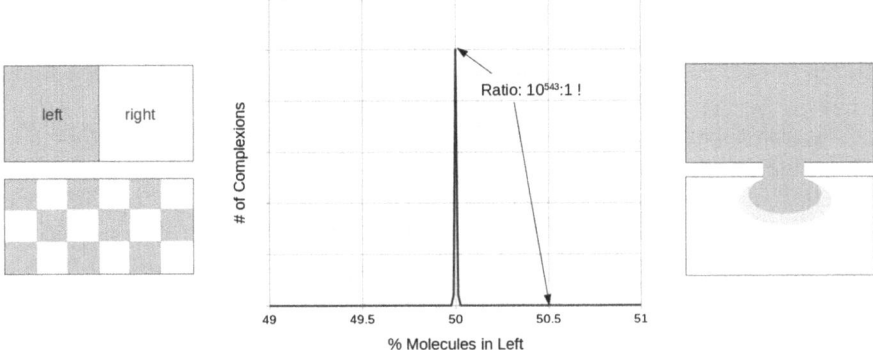

Fig. 2.1 Equilibrium distribution of atoms

in a real world situation, the state with the most complexions dominates to an almost unbelievable degree.

Consider 100 million gas molecules distributed between two equal volumes, labeled left and right in Fig. 2.1. Each molecule could be in either the left hand volume or the right hand volume. So at this macroscopic level of description there is a total of 2^N different spatial complexions. A particular macroscopic distribution would be L molecules in the left-hand volume, and $R = N - L$ molecules in the right-hand volume. The number of spatial complexions in this macroscopic state is the different number of ways of splitting N molecules into two groups with L in one group, R in the other. This is given by the binomial distribution

$$W(L) = \frac{N!}{L!R!} \tag{2.5}$$

Here $N! = N \times (N-1) \times (N-2) \ldots \times 1$, etc. are factorial functions. A graph of $W(L)$ is shown in the central panel of Fig. 2.1. The abscissa gives the distribution in terms of the percentage of molecules in the left hand volume. So 50% corresponds to an equal or 50:50 distribution. Note that the distribution is so narrowly peaked around $L = R$ that the abscissa scale is greatly expanded, with a range of just 49–51%. The real story comes by putting the numbers into the binomial formula. If we move away from the equi-distribution peak at 50:50 by just a small amount, 0.25%, the number of complexions falls by a staggering factor of $1/10^{543}$. Furthermore, if we calculate what fraction of all 2^N complexions is contained in the narrow range of distributions from 49.98:50.02 to 50.02:49.98 spanning this peak, we find that it contains *more than* 99.993% *of all complexions.*

In other words *almost every* complexion is indistinguishable from the most likely, 50:50, state. The most likely state, plus all those states that are macroscopically *indistinguishable* from it, account for all but a very very very tiny fraction of all complexions. Since all complexions are equally likely, we will almost never stray from one of the typical equilibrium-like complexions. The percentages in this example are for 100 million molecules, which is a rather small amount of matter, about 5 million billionths of a gram of oxygen gas. For truly macroscopic amounts of matter the complexions will cluster even more closely around the maximum entropy 50:50 state.

It is also important to understand what Boltzmann's postulate is *not* saying. It is not saying that the macroscopic behavior is a result of the atoms exploring all the complexions (the so-called ergodic hypothesis), or even most of the complexions. Even for this small system of 100 million atoms and at this coarse level of description the total number of complexions is about 1 followed by 30 million zeros. The system can only explore a tiny fraction of the possible complexions in the time we observe it. Therefore it will almost never encounter the vanishing minority of complexions that are detectably unlike the most probable or maximum entropy states.

2.3   The Maximum Entropy State for Gas, Mixtures and Solutes Is Uniform Density

The partition of the volume in Fig. 2.1 into left and right regions is arbitrary. Another equi-volume partition is illustrated by the checkerboard pattern in the lower left of the figure. The same argument about distributions applies to any equi-volume division: almost all complexions will correspond to an equal distribution. From this we conclude that in the absence of external forces the maximum entropy distribution, the one that can happen the most ways, the one that represents all those that will be observed at equilibrium, is a uniform distribution of gas in the available volume.

Now let the gas initially be confined to one volume. Then a valve is opened to a second, equal volume, as in the right panel of the figure. As the gas molecules move from one complexion to another because of their thermal motion, over time they are overwhelmingly likely to end up in the equilibrium-like complexions, the ones with the gas equally distributed between the volumes, simply because almost every complexion is like this. So the gas expands to uniformly occupy all the volume. This is not because the second volume has a vacuum—in fact the expansion will occur even if the second volume is filled with a different gas. The mixing of the two gases is just a by-product of each gas moving independently to states for which there are the most complexions. A similar argument applies to mixing of liquids: A drop of ink mixing with water, or milk stirred into coffee. The molecules of ink or milk are simply moving to complexions belonging to the most likely distribution—a uniform distribution in their available volume. Similarly with any solute dissolved in a solvent. The counting of solute spatial complexions is analogous to that of a gas in the same volume as the solution. If there are no strong interactions between the solutes, and the solute concentration is not too high, then the most likely distribution is a uniform density.

To put in some numbers, consider $10\,\text{cm}^3$ of gas at a pressure of 1 atm and a temperature of 273 K. There are about $N = 2.4 \times 10^{20}$ molecules of gas (using a molar volume of 24.6 L, and Avogadro's number, 6.02×10^{23}). If this gas is allowed to expand into twice the volume, the number of complexions goes up by a factor of $2^N \approx 10^{10^{20}}$, which is 1 followed by about 100 quintillion zeros. That gas is not going to spontaneously go back into the smaller volume any time soon.

2.4 The Entropy of an Ideal Gas

An ideal gas is one in which the molecules have no volume, and make no interactions. The molecules are basically modelled as points, whose only properties are mass, position and velocity. An ideal gas is also a good model for gases like nitrogen, oxygen and argon at normal temperatures and pressures. For these reasons it is often used to illustrate thermodynamic principles. Since there are no forces, the total number of complexions is given by Eq. 1.21. The number of velocity complexions according to Eq. 1.9 is proportional to $E_K^{3N/2}$ and so it is also proportional to $T^{3N/2}$, since temperature is proportional to the total kinetic energy for a fixed number of atoms. For the spatial complexions, we imagine the volume V occupied by the gas to be divided up into a large number of small cubic regions of equal volume b. There are V/b ways of placing a single molecule into one of the small boxes. In an ideal gas the molecules are all independent of each other so there are $(V/b)^2$ ways of placing two molecules into the boxes, and so on. Thus the total number of spatial complexions is proportional to V^N. Clearly the absolute number of spatial complexions depends on how finely we divide space, but for any process of interest involving a volume change, the entropy change will depend on the ratio of the numbers of complexions, and b will cancel out. Putting the spatial and velocity complexion results together, the total number of complexions for an ideal gas is

$$W \propto V^N T^{3N/2} \tag{2.6}$$

The ideal gas entropy is then

$$S_{id} = Nk_b \ln V + \frac{3}{2} Nk_b \ln T + C \tag{2.7}$$

Note the spatial and velocity contributions to the entropy are additive, which follows from what was said earlier, that in the absences of forces the spatial and velocity complexions can be realized independently of the other, so the total number of complexions is given by their product (Eq. 1.21). Again, C is a constant of no interest since it does not depend on T or V and so it will cancel out for any change in these two thermodynamic quantities.

2.5   The Spatial Distribution of a Gas Under Gravity: The Barometric Equation

One of Boltzmann's very first applications of the distribution that now bears his name was to the distribution of a gas in a gravitational field (Boltzmann 1877). Taking an altitude of zero (mean sea level) as a reference, the potential energy of a gas molecule of mass m at a height h above sea level is mgh. Then Eq. 1.24 gives the probability of that gas molecule being at height h as

$$p(h) \propto exp\left(\frac{-mgh}{k_b T}\right). \tag{2.8}$$

Consequently the density (and pressure) of gas has the distribution

$$\rho(h) = \rho(0)\, exp\left(\frac{-mgh}{k_b T}\right) \tag{2.9}$$

where $\rho(0)$ is the density at sea level. Using an approximate molecular weight of 30 g/mole for atmospheric gas, the mass per molecule is 5×10^{-26} kg, $g = 10$ N/kg. With $T = 300$ K and $k_b = 1.38 \times 10^{-23}$ J/K, the factor $k_b T/mg$ has the value 8280 m. In other words, the equation gives an atmospheric density half that of sea level at about 5700 m, or 70% of the way up Mt. Everest. Here we are not correcting for the decrease in temperature with altitude. Equation 2.9 is known as the barometric equation. It gives the maximum entropy arrangement of gas molecules in the atmosphere at constant temperature. This demonstrates that the principle of maximum entropy can operate on a very large scale. Note that on the laboratory scale, these variations in density with height are tiny, so the uniform distribution of Sect. 2.3 is still the maximum entropy distribution.

2.6   The Distribution of Velocities at Equilibrium

Sections 2.3 and 2.5 give the maximum entropy spatial distributions of a gas under different conditions. We now consider the maximum entropy distribution of kinetic energy in a gas. As illustrated in Fig. 1.4, for a fixed total kinetic energy E_K all the possible velocity complexions can be represented by a $3N$-dimensional vector of length $R = \sqrt{2E_K/m}$, $3N$ being the number of velocity terms. Since these velocity vectors are of constant length R they trace out the $(3N - 1)$-dimensional surface of a $3N$-sphere, the number of degrees of freedom being reduced by one because of the constant energy constraint. We can use the velocity-sphere concept to describe the equilibrium distribution of kinetic energy/velocity components in more detail.

First, we know from Sect. 1.7 on heat flow that in a region containing a large number of atoms the temperature should be uniform, i.e. on average the kinetic energy is uniformly distributed. However this does not mean that at the atomic level at any particular moment all the atoms have exactly the same kinetic energy, distributed equally among the x, y and z components. In fact this completely uniform distribution is very unlikely: It corresponds to a complexion where $|u_1| = |\gamma_1| = |w_1| = |u_2| = \ldots |w_N|$. Geometrically, this corresponds to a special direction in 3N-dimensional space for the velocity vector, namely aligned exactly on one of the $3N$ diagonals. But there many more non-diagonal directions in the 3N-dimensional sphere to which the velocity vector could point. Thus there will be a range of velocities among the atoms. The form of this velocity distribution can be obtained from the velocity-sphere, as follows.

Consider any component of the velocity of any atom, say the x component of the ith atom. If the distribution of velocity vectors is uniform over the $(3N - 1)$-dimensional surface of the velocity sphere, as per Boltzmann's postulate, then what is the probability distribution of this velocity component? If this velocity component has a particular value u_i, where $-R \leqslant u_i \leqslant +R$, and $R = \sqrt{2E_K/m}$ is the radius of the velocity sphere, then this constrains all possible 3N-dimensional velocity vectors to trace a $(3N - 2)$-dimensional 'circle' C of radius

$$R_c = \sqrt{R^2 - u_i^2} = R\left(1 - \frac{\frac{1}{2}mu_i^2}{E_K}\right)^{1/2} \tag{2.10}$$

on the $(3N - 1)$-surface, see Fig. 2.2. In Eq. 2.10 $\frac{1}{2}mu_i^2$ is the x component of the kinetic energy of atom i.

Technically, in high dimensions the $(3N - 1)$-dimensional surface of a 3N-sphere and a $(3N - 2)$-dimensional circle on this surface are both hyperspheres, but retaining the surface/circle terminology for the minus-one and minus-two dimensions makes the explanation easier.

Fig. 2.2 Possible velocities when the total kinetic energy E_K and one velocity component u_i are fixed

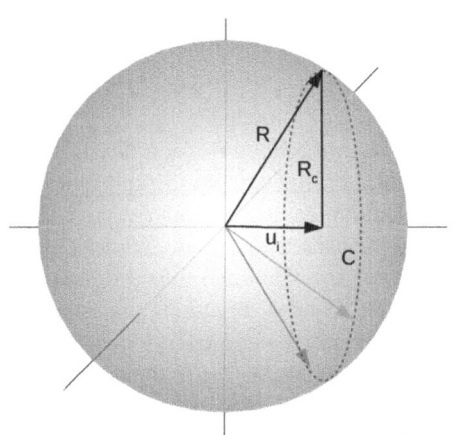

The number of possible velocity complexions for fixed E_K and u_i is proportional to the 'circumference' l_C of the hyper-circle C, which in turn depends on the $(3N - 2)$th power of its radius, so

$$l_C \propto R^{3N-2} \left(1 - \frac{\frac{1}{2}mu_i^2}{E_K}\right)^{(3N-2)/2} \propto \left(1 - \frac{\frac{1}{2}mu_i^2}{E_K}\right)^{3N/2} \tag{2.11}$$

where like before we make the approximation $3N - 2 \approx 3N$ since N is very large, to make the equations tidier. We have also dropped the R term. It is constant because the total kinetic energy, E_K, is constant. By Boltzmann's postulate, the probability of a particular value of u_i is proportional to the total number of velocity complexions with this value,

$$p(u_i) \propto \left(1 - \frac{\frac{1}{2}mu_i^2}{E_K}\right)^{3N/2}. \tag{2.12}$$

Re-writing the right hand side using the identity $x = e^{\ln x}$ we have

$$p(u_i) \propto exp\left(\frac{3N}{2}\ln(1 - \frac{\frac{1}{2}mu_i^2}{E_K})\right). \tag{2.13}$$

Since there are very many atoms among which to partition the total kinetic energy, we can assume that almost all the time a particular atom has only a small fraction of the total, so $\frac{1}{2}mu_i^2 << E_K$ and we can use the logarithm approximation $\ln(1 - x) \approx -x$ for small x to obtain

$$p(u_i) \propto exp\left(-\frac{\frac{1}{2}mu_i^2}{\frac{2}{3}E_K/N}\right) \propto exp\left(-\frac{\frac{1}{2}mu_i^2}{k_bT}\right) \tag{2.14}$$

using $k_bT = \frac{2}{3}\hat{E}_K$ again. By symmetry, each of the $3N$ velocity components has the same probability distribution as Eq. 2.14. This is the famous Maxwell-Boltzmann distribution for velocities, first derived for gases, but present also in any solid or liquid at equilibrium. We see it at equilibrium simply because it is the distribution that can happen the most ways. It is the maximum entropy distribution. What it means is that at any instant some atoms will have more kinetic energy than others. The probability of an atom having a certain amount of kinetic energy in any of the x, y and z directions decreases exponentially with increasing energy. The distribution of kinetic energy components described by Eq. 2.14 has a number of important consequences in the macroscopic world.

2.6.1 Evaporative Cooling

Consider evaporation of water. This is a process by which some molecules of water escape from the liquid and become a gas. This takes energy, because the molecules in the liquid are making strong hydrogen bonding interactions with each other. Therefore the molecules most likely to evaporate are the ones with higher kinetic energy, at the upper end of the Maxwell-Boltzmann distribution (Fig. 2.3). Preferential escape of the higher kinetic energy molecules means that the average kinetic energy of the remaining molecules is lower, i.e. evaporation cools a liquid. Evaporative cooling is an important phenomenon for weather, thermo-regulation in living things and in industrial processes. It would not occur if all the molecules had exactly the same kinetic energy, so we can say that it is one of the consequences of the maximization of entropy.

2.6.2 Chemical Reactions

Chemical and biochemical reactions of all kinds start with the reactants interacting with each other to form some kind of reactive complex, then this complex undergoes some kind of rearrangement of atoms, followed by breakdown of the complex into the products. Rearrangement of the atoms during the reaction most often involves going over some potential energy barrier. As with evaporation, the molecules that are most likely to react are the ones whose atoms have higher kinetic energy, those at the upper end of the Maxwell-Boltzmann distribution. Figure 2.3 shows how this distribution changes with temperature, in this case going from 300 to 360 K, a 20% increase. We know that the average kinetic energy increases proportionally, because

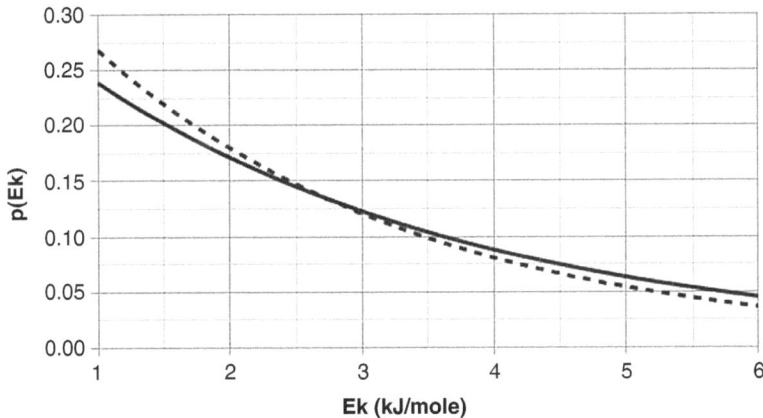

Fig. 2.3 Probability of kinetic energy component values at 300 K (dotted line) and 360 K (solid line)

of the relation $k_b T = \frac{2}{3}\hat{E}_K$. However, as the factor $k_b T$ occurs in the exponent of Eq. 2.14, the distribution is flatter at higher T. From the figure one can see that in fact the number of atoms with lower than average kinetic energy actually decreases. This means that the number of atoms with higher kinetic energy increases more than does the average. In this example, the probability of an atom having more than four times the average kinetic energy increases by 60%, those with more than eight times the average are thrice as likely. The end result is that chemical and biochemical reactions are very sensitive to temperature. As a rule of thumb, a $10\,°C$ increase can double the rate. This has important implications for chemistry and biology. Again, this sensitivity comes from the properties of the maximum-entropy Maxwell-Boltzmann distribution.

Summary
Clausius' definition of an entropy change as the transferred heat divided by the temperature is equivalent to the logarithm of the change in number of kinetic energy complexions, the number of ways to distribute heat. More generally the entropy of a state is a logarithmic measure of the total number of complexions, kinetic plus spatial, available to that state. Equilibrium corresponds to the maximum entropy state, the state with the most complexions. This leads in the absence of any constraints to a uniform distributions of gas, mixtures and solutes in the accessible volume, to the exponential dependence of atmospheric pressure with altitude, and to the exponential distribution of atomic velocities (The Maxwell-Boltzmann distribution).

Chapter 3
Entropy and Free Energy

Abstract In thermodynamics a quantity known as the free energy plays a central role. Changes in free energy determine the direction of chemical reactions and the values of equilibrium constants. In this chapter it is shown that a free energy change is just a different way of expressing the total entropy change, and that it consists of two distinct entropic components, spatial and kinetic.

Keywords Entropy · Free energy · Enthalpy · Equilibrium constant · Partition function · Gibbs-Einstein equation

3.1 Separating Entropy into Spatial and Velocity Components

Applying the relation $S = k_b ln W + C$ to the general expression for the total number of complexions, Eq. 1.31, we obtain the most general expression for the entropy, valid whether there are forces or not:

$$S = k_b ln Z + Constant \tag{3.1}$$

where Z is the partition function. For an entropy difference between states A and B:

$$\Delta S = k_b ln(Z_B) - k_b ln(Z_A) = k_b \Delta ln Z. \tag{3.2}$$

Now the partition function is

$$Z = \sum_i^{W_s} exp(-U_i / k_b T) \tag{3.3}$$

which is a weighted sum over spatial complexions, where each spatial complexion is weighted by the number of kinetic energy complexions available to it. So the spatial and kinetic energy contributions to entropy seem to be combined in a complicated way. However it is straightforward to separate any change in total entropy going from state A to state B, given by Eq. 3.2, into the contributions from spatial complexions, ΔS_s and from kinetic energy complexions, ΔS_v as follows:

1. The average potential energies in states A and B can be obtained from Eq. 1.28 as:

$$U_A = \frac{1}{Z_A} \sum_{i}^{W_s(A)} U_i exp(-U_i/k_b T), \quad U_B = \frac{1}{Z_B} \sum_{i}^{W_s(B)} U_i exp(-U_i/k_b T).$$

(3.4)

The change in average potential energy is $\Delta U = U_B - U_A$. Applying conservation of energy the change in kinetic energy is $\Delta Q = -\Delta U$.

2. Equation 1.15 is used to calculate the change in number of kinetic energy complexions:

$$\frac{W_v(B)}{W_v(A)} = exp(-\Delta U/k_b T),$$

(3.5)

assuming again that the number of atoms N is large enough that $\Delta Q << E_K$, i.e. the temperature change is negligible

We then take the logarithm to obtain the kinetic energy contribution to entropy

$$\Delta S_v = -\Delta U/T$$

(3.6)

The spatial contribution to the entropy is then obtained by subtracting the kinetic energy contribution from the total

$$\Delta S_s = \Delta S - \Delta S_v = k_b \Delta ln Z + \Delta U/T.$$

(3.7)

Put another way, the total entropy change can be cleanly separated into its two contributions

$$k_b \Delta ln Z = \Delta S = -\Delta U/T + \Delta S_s$$

(3.8)

where $\Delta S_v = -\Delta U/T$, and ΔS_s, are the kinetic complexion and spatial complexion contributions to the entropy change, respectively.

3.1.1 The Gibbs-Einstein Equation for Entropy

There is an alternative approach to separating the total entropy into spatial and kinetic components. It is less direct, but it leads to the third fundamental equation for entropy. We can write as an identity

$$- k_b \sum_{i=1}^{W_s} p_i ln p_i = k_b \sum_{i=1}^{W_s} p_i (ln Z + U i / k_b T) \tag{3.9}$$

where Eq. 1.25 has been used to substitute for $ln p_i$. The first term sums to $k_b ln Z$ since $ln Z$ can be brought in front of the summation and $\sum p_1 = 1$. Using Eq. 1.28 the second term simply gives the average energy divided by temperature, U/T. So

$$- k_b \sum_{i=1}^{W_s} p_i ln p_i = k_b ln Z + U/T. \tag{3.10}$$

Comparison with Eq. 3.7 shows that

$$- k_b \sum_{i=1}^{W_s} p_i ln p_i = S_s + C. \tag{3.11}$$

This is known as the Gibbs-Planck-Einstein expression for the spatial entropy. The constant C is unimportant since we are only ever interested in changes in entropy, wherein it cancels.

3.2 Entropy, Enthalpy and Work

The final thing we have to account for is the possibility that, when there is a change in the potential energy due to a change in the spatial complexions of the atoms, not all that energy turns into kinetic energy. Some of that energy may be used to do work w. Quite simply, the amount of heat produced is reduced by the amount of work done: $\Delta Q = -\Delta U - w$. A common example is when there is a change in volume, ΔV, which does work $w = P \Delta V$ against an external pressure P. The change in kinetic energy is now $-\Delta U - P \Delta V$ and Eq. 3.8 for the change in total entropy is replaced by

$$\Delta S = k_b \Delta ln Z = -\Delta U/T - P \Delta V/T + \Delta S_s \tag{3.12}$$

Other types of work include surface free energy work $w = \gamma \Delta A$ done by a change ΔA of surface area with free energy (surface tension) of γ, electrical work $w = q \Delta \Psi$ caused by a charge q moving across a voltage difference $\Delta \Psi$, and magnetic work $w = \Delta(\mathbf{B} \cdot \mathbf{m})$, where $\mathbf{B}$ and $\mathbf{m}$ are the magnetic field and dipole moment, respectively.

3.2.1 Connecting the Language of Entropy to the Language of Energetics

For historical reasons having to do with the importance of calorimetry (measurement of heat changes) in the development of thermodynamics, and the discovery of the first law of thermodynamics before the second law, Eqs. 3.8 and 3.12 for the total entropy change are never presented in this form. Instead, they are written in what might be described as the language of energetics. To translate, simply multiply by $-T$ (Biot 1955), giving

$$\Delta A = -k_b T \Delta \ln Z = \Delta U - T \Delta S_s \qquad (3.13)$$

if there is no volume change, and

$$\Delta G = -k_b T \Delta \ln Z = \Delta U + P \Delta V - T \Delta S_s \qquad (3.14)$$

if there is a volume change. $\Delta H = \Delta U + P \Delta V$ is called the enthalpy change. A and G are known as the Helmholtz and Gibbs free energies, and apply to experimental conditions of constant volume or constant pressure, respectively. The heat change, either ΔU or ΔH, depending on conditions, can be measured by calorimetry.

When interpreting the physical meaning of Eqs. 3.13 and 3.14, the introduction of the terminology 'free energy' and the negative sign that operates on the entropy change are unfortunate historical accidents (Craig 2005). Each of the terms on the right hand side of these equations describe an entropy contribution. So the statement that the minimization of free energy is the condition for equilibrium obscures the fact that the condition is actually a maximization of total entropy, a maximization of the combination of spatial and kinetic energy complexions. Another misconception is to describe the ΔS_s term as the change in entropy of the 'system of interest' and the ΔU or ΔH term as representing the change in entropy of the 'surroundings' or the 'environment'. The two terms do not represent a division in that spatial sense, but a division between the spatial and kinetic contributions to entropy. The same two contributions Boltzmann identified in his original work (Boltzmann 1877).

3.3 Equilibrium Constants and Entropy

One of the most useful thermodynamic quantities is the equilibrium constant. Examples of equilibrium constant include the partition coefficient of a solute between two solvents, the equilibrium between reactants and products in a chemical reaction, molecular conformation equilibrium constants and binding constants. To be concrete, consider the case of a conformational equilibrium.

Suppose that our collection of N atoms contains one molecule of X which is made up of N_X atoms, where $N_X \ll N$. Further, suppose that X can adopt just two conformations, A and B. This means that we can divide the set of all possible spatial complexions of the N atoms into two subsets: one subset has the atoms of X arranged in conformation A, the other subset has X arranged in conformation B. The non-X atoms are free to adopt any spatial complexions consistent with each conformation of X. Boltzmann's fundamental postulate says that the relative probabilities of finding X in conformation A *versus* B will be equal to the ratio of the total number of complexions possible with each conformation:

$$K_{AB} \equiv \frac{p_A}{p_B} = \frac{W(A)}{W(B)} \tag{3.15}$$

where the total number of complexions $W = W(A) + W(B)$ is determined by the partition function Z as before. This ratio of probabilities defines the equilibrium constant K_{AB} between A and B. Determination of $W(A)$ and $W(B)$ amounts to splitting the partition function sum over all spatial complexions in Eq. 1.31 into two parts, namely

$$W(A) = C' \sum_{i=\{A\}} exp(-U_i/k_bT) \qquad W(B) = C' \sum_{j=\{B\}} exp(-U_j/k_bT) \tag{3.16}$$

where the partial sums are over all spatial complexions where the N_X atoms of X are either in the A or B conformations, respectively. Then the equilibrium constant can be written as

$$K_{AB} = \frac{\sum_{i=\{A\}} exp(-U_i/k_bT)}{\sum_{j=\{B\}} exp(-U_j/k_bT)} = \frac{Z_A}{Z_B} \tag{3.17}$$

which is the ratio of the partition functions for the A and B conformations, respectively. If we use the identity

$$Z = exp\left(\frac{k_bT \ln Z}{k_bT}\right) \tag{3.18}$$

(see Appendix) we can express the equilibrium constant at constant pressure as

$$K_{AB} = exp\left(\frac{(k_bT \ln Z_A - k_bT \ln Z_B)}{k_bT}\right) = exp(-\Delta G_{AB}/k_bT). \tag{3.19}$$

This important thermodynamic relationship connects the experimentally measurable equilibrium constant K_{AB} to the free energy difference, ΔG_{AB}. Noting the term $\Delta G_{AB}/-T$ in the exponent of Eq. 3.19 it is more informative to say that the equilibrium constant is determined by the difference in total entropy (spatial and kinetic) between the two conformations:

$$K_{AB} = exp(\Delta S_{AB}/k_b). \tag{3.20}$$

The equilibrium will favor the conformation that has the greatest total entropy (number of complexions).

Equation 3.17 was obtained for a system containing a single molecule of X, in other words the equilibrium constant would refer to the average occupancy of the A and B conformations over a period of time long compared to the time X spends in each conformation. Alternatively, if the system contained a macroscopic number of X molecules, at any instant the ratio of the number in the A and B state (or in solution the ratio of concentrations) will be given by the same probability ratio:

$$\frac{N_A}{N_B} = \frac{[A]}{[B]} = \frac{p_A}{p_B} = K_{AB} \tag{3.21}$$

where it is assumed that the concentration of X is low enough that there is no interaction between different molecules of X. In other words each molecule of X independently follows the probability distribution (Eq. 3.15) derived for a single molecule. This is the usual dilute solution or ideal solute assumption.

More generally, for any equilibrium one can partition the spatial complexions into two sets, each corresponding to states/chemical species on one side of the equilibrium. One then evaluates the free energy of each side by forming the partition function sums over each of the two disjoint subsets. So for example, to evaluate the equilibrium constant for the chemical reaction $X + Y \rightleftharpoons XY$, in Eq. 3.17 the 'state A' sum would be over all spatial complexions corresponding to non-bonded X and Y, while the 'state B' sum would be over all spatial complexions where X and Y are bonded.

Summary
1. The partition function Z is the sum of the Boltzmann factors over the spatial complexions.

$$Z = \sum_{i=1}^{W_s} exp(-U_i/k_b T).$$

  It is proportional to the total number of complexions
2. The change in total entropy is given by the change in the logarithm of the partition function

$$\Delta S = k_b \Delta ln Z$$

3. The change in the kinetic energy (atomic velocity) contribution to the entropy is obtained from the change in mean potential energy as

(continued)

$$\Delta S_v = \frac{\Delta Q}{T} = -\frac{\Delta U}{T}$$

4. The change in the spatial contribution to the entropy is given by the difference of (2) and (3), or directly from the Gibbs-Planck-Einstein expression

$$\Delta S_s = k_b \Delta ln Z + \Delta U/T = \Delta \left(-k_b \sum_{i=1}^{W_s} p_i ln p_i \right)$$

The importance of the partition function Z can now be appreciated, despite its inauspicious debut as a normalization constant for the probability of spatial complexions. The logarithm of the partition function gives, to within an additive constant, the total entropy due to both spatial and kinetic energy complexions.

Chapter 4
Entropic Forces

Abstract This chapter examines the origin of entropic forces. These forces are compared with forces that arise from interaction potentials, like gravitational and electrical forces.

Keywords Force · Entropic force · Entropy derivative · Entropic work · Potential energy

4.1 Potential Energy and Forces

In physics fundamental forces such as gravitational and electrical forces can be derived from their corresponding interaction potentials. For example the potential energy of two atoms of masses m_1 and m_2 separated by a distance r_{12} due to their gravitational interaction is

$$U = \frac{Gm_1m_2}{r_{12}} \tag{4.1}$$

where G is the gravitational constant. The force acting on atom 1 due to atom 2 is given by the gradient of this potential with respect to the position of atom 1:

$$f_1 = -\frac{dU}{dr_i} \tag{4.2}$$

which gives Newton's familiar inverse square force law:

$$f_i = \frac{Gm_1m_2}{r_{12}^2} \, . \tag{4.3}$$

The force is directed along the line between the two atoms in the direction of decreasing potential. The force on atom 2 is obtained in the same manner and is equal and opposite.

Another familiar example is a metal spring. For small displacements the spring potential is

$$U = \frac{1}{2}Kx^2 \qquad (4.4)$$

where K is the spring constant, and x is the displacement of the spring away from its unstrained length. The gradient of this potential gives the restoring force

$$f = -Kx \qquad (4.5)$$

which is linear in the displacement (Hooke's Law). The minus sign indicates that the force is in the opposite direction to the displacement and acts to restore the spring to its unstrained state.

4.2   Entropic Forces in a Gas

The interaction potential (IP) derived gravitational and electrical forces act on individual atoms but at the macroscopic level they are very concrete: We can feel the force exerted by a weight held in our hand. We can feel the resistance when we stretch a metal spring: It arises from electrical forces bonding the metal atoms together as we slightly displace the atoms from their equilibrium separations. Forces also arise from entropic effects, but rather than coming from individual atom-atom interactions they arise from the collective behavior of atoms. Because of this they perhaps seem less concrete; but they are just as real and tangible. The ideal gas again forms a good starting point.

4.2.1   Ideal Gas Pressure

Consider a cylinder fitted with a piston and filled with N molecules of an ideal gas, Fig. 4.1. The area of the piston is A, and the position of the piston relative to the bottom of the cylinder is h. The volume of the gas is $V = Ah$, and its pressure is given by Boyles Law, Eq. A.20 as

$$P = Nk_bT/V . \qquad (4.6)$$

The force acting on the piston from the gas is

$$f = PA = Nk_bT/h. \qquad (4.7)$$

Fig. 4.1 Cylindrical piston filled with an ideal gas

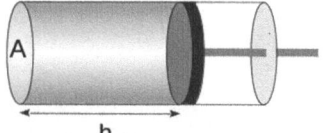

The entropy of the gas, from Eq. 2.7, is

$$S_{id} = Nk_b ln Ah + \frac{3}{2} Nk_b ln T + C$$

where C is an unimportant constant. Following the rule discussed in Sect. 3.2.1 for translating from entropy to the energeticist's terminology, let us define for any S the corresponding entropic potential, $-TS$ (Biot 1955). This has units of energy. For the gas in the cylinder, its entropic potential is

$$-TS_{id} = -Nk_b T ln Ah - \frac{3}{2} Nk_b T ln T - TC.$$

We find, in exact correspondence with Eq. 4.2, that the negative gradient of this entropic potential with respect to the piston position gives a force

$$-\frac{d(-TS_{id})}{dh} = \frac{Nk_b T}{h}$$

which is identical to that obtained from Boyle's law. The term involving $ln T$ and the constant term don't depend on the piston position and disappear. Note that the direction of an entropic force is such as to increase the entropy.

In an ideal gas there are no forces between the atoms. Instead, the force exerted by an ideal gas is entirely entropic in origin, as demonstrated by the fact that it can be derived from Eq. 2.7 for the total entropy. The signature of an entropic force is that it scales linearly with the temperature. IP-derived forces, in contrast, are temperature invariant. At least, the *form* of the force is T-invariant, although the positions of atoms may vary with temperature producing a trivial temperature dependence. What this difference in behavior with respect to temperature means is that as T is increased, entropic forces will eventually overwhelm IP-forces.

4.2.2 The Gas Cylinder as an Entropic Spring

Consider now our cylinder of ideal gas when the piston is loaded with a weight of mass m, as shown in Fig. 4.2. We assume that the mass of the piston is negligible. With no load the pressure in the cylinder must equal atmospheric pressure, P_{atm}. The equilibrium position of the piston is then given by

$$V_0 = \frac{nRT}{P_{atm}} \tag{4.8}$$

where V_0 is the volume of the gas, and n is the number of moles. When the piston is loaded with the weight and comes to its new equilibrium position the pressure in the cylinder increases to $P_{atm} + mg/A$ to counteract the external atmosphere and

Fig. 4.2 Gas spring. Left: piston at equilibrium at the zero point. Right: piston is displaced downward (in the positive direction) a distance x by a weight of mass m

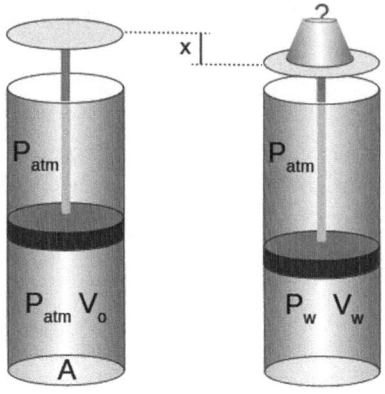

the weight. The new equilibrium position in terms of the gas volume is

$$V_w = \frac{nRT}{P_{atm} + mg/A} \tag{4.9}$$

and the change in volume is $\Delta V = V_w - V_0$, which is negative.

We now work the problem in terms of entropy changes. The change in gravitational potential energy is $\Delta U_g = -mgx = mg\Delta V/A$ and the corresponding change in entropy (kinetic) is

$$\Delta S_v = \frac{-mg\Delta V}{AT} \tag{4.10}$$

which is positive and linear in displacement ΔV. The change in spatial entropy from compression of the gas in the cylinder is obtained from the ideal gas entropy equation as

$$\Delta S_s = nRln\frac{V_w}{V_0} = nRln\left(\frac{V_0 + \Delta V}{V_0}\right) \tag{4.11}$$

which is negative and non-linear in displacement.

Now when the piston moves down and compresses the gas in the cylinder by ΔV, the atmosphere above the piston *expands* by $-\Delta V$, and the entropy change of the atmosphere must be included:

$$\Delta S_{atm} = n_{atm}Rln\left(\frac{V_{atm} - \Delta V}{V_{atm}}\right)$$

$$= n_{atm}Rln\left(1 - \frac{\Delta V}{V_{atm}}\right)$$

$$\approx -\frac{n_{atm}R\Delta V}{V_{atm}}$$

since $V_{atm} >> \Delta V$. We don't need to know the number of moles or the volume of the atmosphere, since by the ideal gas law their ratio is given by $n_{atm}R/V_{atm} = P_{atm}/T$, which finally gives the change in entropy of the atmosphere as

$$\Delta S_{atm} = -\frac{P_{atm}\Delta V}{T} \tag{4.12}$$

which is linear in displacement and positive. The total change in entropy is

$$\Delta S_{tot} = -\frac{mg\Delta V}{AT} - \frac{P_{atm}\Delta V}{T} + nRln\left(\frac{V_0 + \Delta V}{V_0}\right) \tag{4.13}$$

To summarize, since ΔV is negative, the first two terms are positive and the third term is negative. The first term is the change in the kinetic part of the entropy, while the second and third terms give the spatial contribution, including both gas in the cylinder and in the atmosphere. The maximum in entropy is obtained by taking the derivative with respect to the displacement ΔV and setting it equal to zero:

$$0 = -\frac{mg}{AT} - \frac{P_{atm}}{T} + \frac{nR}{(V_0 + \Delta V)} \tag{4.14}$$

Multiplying through by T and solving for $V_w = V_0 + \Delta V$ we get the piston position of maximum entropy,

$$V_w = \frac{nRT}{P_{atm} + mg/A}, \tag{4.15}$$

which is the same as that given by force balance, Eq. 4.9. Figure 4.3 plots the various entropy terms against linear displacement $x = \Delta V/A$ with parameters $T = 300\,K, P = 1\,atm, n = 0.1$ moles, $A = 0.01\,m^2$. Loading with a mass of

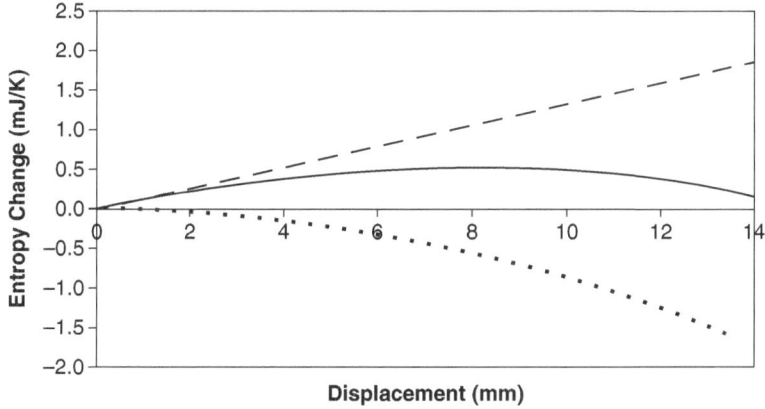

Fig. 4.3 Entropy contribution from gravitational potential energy change plus external atmosphere (dashes), gas in the cylinder (dots) and total entropy (solid line)

3.3 kg gives an equilibrium displacement downwards of 7.8 mm, and a maximum entropy change of 0.43 mJ/K. This corresponds to an increase in the number of complexions by a factor of about $10^{1.4 \times 10^{19}}$.

Returning to the total entropy we can write Eq. 4.13 as

$$\Delta S_{tot} = -\frac{\Delta U}{T} - \frac{P_{atm} \Delta V}{T} + \Delta S_s \tag{4.16}$$

and translating this into energeticist's terms by multiplying by $-T$ we get

$$\Delta G = \Delta U + P_{atm} \Delta V - T \Delta S_s = \Delta H - T \Delta S_s \tag{4.17}$$

where $\Delta H = \Delta U + P_{atm} \Delta V$ is called the change in enthalpy. We leave the subscript s on the entropy term to remind us that this is just the change in spatial entropy of our system (the gas in the cylinder). The minimum free energy condition for equilibrium is seen to be formally equivalent to the maximum total entropy condition (kinetic plus spatial entropy). The equivalent conditions for equilibrium are maximum entropy, force balance and minimum free energy. Since there is a change in volume of our system (gas cylinder/piston) the enthalpy rather than the change in potential energy appears in the free energy expression.

4.2.3 Osmotic Pressure

An important force in living organisms is the osmotic pressure. This too is an entropic force, closely analogous to the ideal gas pressure. If we have a solute that dissolves well in a solvent and which acts as an ideal solute (does not interact with itself or other solutes in the solution) then this solute produces an osmotic pressure given by the equivalent of Boyle's Law:

$$P = mRT \tag{4.18}$$

where m is the solute molar concentration, conventionally expressed in units of moles/liter, and $R = 8.314$ J/K/mole is the gas constant. The osmotic pressure becomes a tangible pressure if the solute is confined by a semi-permeable membrane: a membrane that is permeable to the solvent but not the solute (Fig. 4.4). In this case the solvent, which can move freely through the membrane, plays no role in the production of force other than to keep the solute dispersed. The spatial entropy of the solute is exactly that of the same number of molecules of ideal gas contained in the volume inside the semi-permeable membrane, and so the entropic force on the membrane follows that from the ideal gas law.

Fig. 4.4 Osmotic pressure of
a solute (spheres) exerted on
a semi-permeable membrane
(dotted line)

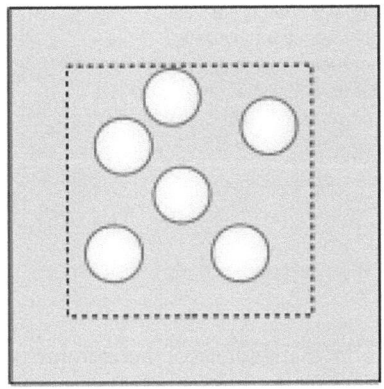

4.3 Entropic Forces in Polymers

Disordered polymers, large molecules that can adopt a multitude of different
conformations, are also the source of entropic forces. We consider the simplest
treatment of these forces to establish the principle: A polymer of N rigid segments
of uniform length b connected by flexible joints (Fig. 4.5). Interactions between
segments are neglected.

The strategy for complexion counting in this case is to model the conformational
states as a three-dimensional random walk. For sufficiently large N the number of
possible chain conformations for a given end-to-end distance r is given by Tanford
(1961)

$$W(r) \propto r^2 exp(-r^2/L^2) \tag{4.19}$$

where $L = \sqrt{2Nb^2/3}$ is the root mean squared (rms) end-to-end distance. Using
Eq. 2.3 the conformational entropy as a function of the end-to-end distance is given
by

$$S_c(r) = k_b(2ln(r) - r^2/L^2) + C. \tag{4.20}$$

C is an unimportant constant since we are only interested in changes with respect to
r. From Eq. 4.20 the polymer's entropy is maximum when its end-to-end distance
is equal to its rms value $r = L$, with a value of $S_c' = k_b(2ln L - 1) + C$. The
maximum entropy state describes the unstrained polymer. If the polymer is stretched
(or compressed) by changing the end-to-end distance by an amount $x = (r - L)$,
then for small extensions ($x << L$) the polymer entropy decreases as the square of
the displacement:

$$S_c(x) = S_c' - 2k_bx^2/L^2 \tag{4.21}$$

Fig. 4.5 Freely jointed
polymer chain in some
random conformation. b is
the monomer length, r is the
end-to-end distance

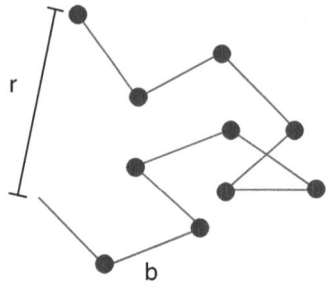

From the gradient of the entropic potential $-T S_c(x)$ with respect to the end-to-end distance the restoring force is

$$f_c = -4k_b T x / L^2, \qquad (4.22)$$

an entropic spring obeying Hooke's law with spring constant $K = 4k_b T / L^2$. The minus sign in Eq. 4.22 indicates that the force always acts to restore the polymer to its unstrained, maximum entropy length of L. Rubber, for instance, consists of cross-linked random polymers, so when two ends of a piece of rubber are pulled the strain is transmitted to individual polymer chains which results in a restoring force on their ends described by Eq. 4.22. In actual rubber there are other contributions to elasticity, but the polymer conformational entropy reduction is the major source of elasticity. Polymer conformational entropy is of course a type of spatial entropy.

4.3.1 A Weight Suspended by a Rubber Cord

By analogy with the gas spring in Sect. 4.2.2 consider a weight of mass m attached to an elastic cord (Fig. 4.6). The cord is made of rubber with a polymer chain density ρ (number of chains per unit volume). The cord in the unstrained state has a length B and a constant cross-sectional area A, so the volume of rubber is AB, and the number of polymer chains is $N_p = \rho A B$. If the weight stretches the band by a distance d the decrease in gravitational potential energy is $\Delta U_g = -mgd$ with a corresponding increase in the kinetic part of the entropy by $\Delta S_v = -\Delta U_g / T = +mgd/T$. The strain induced in the band, as a ratio of its original length, is d/B. For simplicity assume that this strain is homogeneously distributed throughout the rubber. In other words, the strain ratio for each polymer molecule is equal to the macroscopic strain. Hence in Eq. 4.21 $x/L = d/B$. Then the stretching of the band reduces the conformational entropy by

$$\Delta S_c = -2k_b N_p \frac{d^2}{B^2} = \frac{-2k_b \rho A}{B} d^2 . \qquad (4.23)$$

Fig. 4.6 Weight is attached
to an unstrained rubber band
(left), released, and allowed
to come to equilibrium (right)

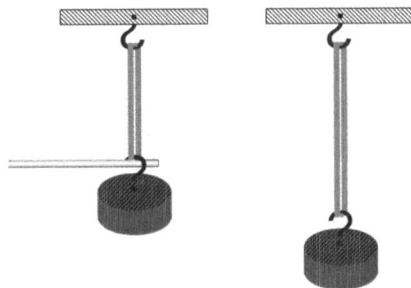

The total entropy change is

$$\Delta S_{tot} = \Delta S_v + \Delta S_c = \frac{mgd}{T} - \frac{2k_b \rho A}{B} d^2 \qquad (4.24)$$

which will be a maximum when the gradient of ΔS_{tot} with respect to d is zero. This
is also the point of gravitational and elastic force balance

$$mg = \left(\frac{4k_b T \rho A}{B} \right) d \, . \qquad (4.25)$$

The chain entropy of the rubber produces a Hooke's law spring with a force constant
that is proportional to the density of the rubber and the cross-sectional area, and
inversely proportional to the length, as one expects. The force constant is also
proportional to temperature, characteristic of entropic forces.[1] The minimum of the
corresponding free energy also occurs at the point of force balance, with a value
given by

$$\Delta G_{tot} = \Delta U_g - T \Delta S_c \, . \qquad (4.26)$$

4.4   Maximum Entropy Is Equivalent to Minimum Free Energy

People who have only encountered free energies at the molecular scale, for example
in the thermodynamics of chemical and biochemical reactions, may be surprised
by their appearance in the simple laboratory scale mechanical examples given here.
Even the distinction between energy and enthalpy contributions (ΔU vs. ΔH) is
here, depending on whether the process is happening at constant volume (the weight
suspended by a rubber cord) or with a volume change at constant pressure (the gas
spring). The free energy is *exactly* the same quantity at both scales. Only the origin
of the potential, and the scale of the system are different. In chemical reactions for

---

[1]The main cause of the space shuttle Challenger disaster was decreased rubber elasticity in the
solid rocket booster O-rings due to the unusually low temperatures before launch.

example, the potential is a quantum mechanical electrostatic potential which is some function of the reactant or product atomic coordinates, and the scale is molecular. But a change in *any* potential energy at any scale always results in a change in heat and therefore a change in the kinetic contribution to entropy. The spatial entropy term just comes from atomic positions, however many atoms there are.

4.5   The Hydrophobic Force

The final entropic force considered here is the hydrophobic force. At the outset it should be said that this is a complex and not yet fully understood force. In part this is because it depends on the properties of liquid water, which is still a somewhat mysterious solvent, and in part because it involves interactions between solutes in water, which by definition takes us into the realm of non-ideal solutions. We do know that the force exists and that it is entropic at room temperature, although there is likely some non-entropic contributions. Also, while the energies involved are quite well known, the range and magnitude of the forces are not. For all these reasons it is often called the hydrophobic effect rather than a force. The reader is referred to two monographs on the subject for more details (Tanford 1980; Ben-Naim 1987).

4.5.1   *Experimental Basis of the Hydrophobic Effect*

It is well known that apolar compounds—liquids like oil, organic solvents like alkanes and benzene, etc. do not mix with water. Shake oil and water together and they rapidly separate into two layers. Apolar solutes in general dissolve poorly in water and if they exceed their solubility limit they will aggregate and come out of solution. The cause of all this is the hydrophobic effect. Consider the case of oil and water. After shaking vigorously, a uniform mixture should be the maximum entropy situation according to argument given in Chap. 2, but instead the two liquids spontaneously de-mix, with an apparent decrease in entropy (Fig. 4.7). This decrease in entropy is only apparent. Thermodynamic measurements show that the water immediately surrounding an apolar solute is in a lower entropy state than bulk water. The entropy of this hydrating water is lowered sufficiently that the mixed state is lower in total entropy than the de-mixed state, hence de-mixing occurs. Put another way, demixing of the oil from water releases this hydrating water into the bulk aqueous phase, with a gain in entropy.

4.5.2   *Mechanism of the Hydrophobic Effect*

It is believed that the hydrating water, because it cannot hydrogen bond with an apolar solute, is forced to form clathrate-like structures around the solute. Clathrates are characterized by pentagonal rings of hydrogen bonded water (Fig. 4.8, Tanford

Fig. 4.7 Pentane and water after shaking (left), and de-mixing (right)

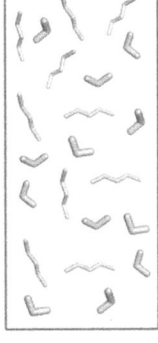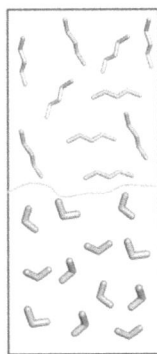

Fig. 4.8 Clathrate water structure around an apolar group (partially hidden, in black)

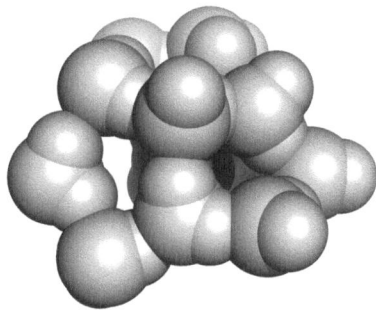

1980). The clathrate is more structured than bulk water. The positions and orientations of clathrate waters are more restricted. This reduces their spatial entropy. Following this line of argument, taking two separated apolar solute molecules and bringing them close enough to exclude water from their interface should result in an increase in total entropy, since the amount of water in contact with apolar surface is decreased. In going thus to a state of higher entropy, somewhere along this pathway there must be a positive gradient in entropy, and hence an attractive hydrophobic force. This force presumably develops once the solute surfaces are close enough exclude water, which is of the order of the diameter of a water molecule, ≈ 0.36 nm. This model predicts that the hydrophobic effect is proportional to the change in apolar surface area exposed to water. This area dependence is borne out quite well by the data (Tanford 1980), which gives the change in water entropy per unit area of about -5.8×10^{-23} J/K/nm^2, which from Eq. 2.2 corresponds to about a 66-fold decrease in water spatial complexions per nm^2. This gives a hydrophobic strength, defined as the free energy change per unit area exposed to water, of 104 J/mole/Å^2. The magnitude of the hydrophobic effect is sufficient for it to play a major role in biology: Helping to determine the structure and stability of proteins and promoting self assembly of biological membranes, for example.

While the structure of water around apolar molecules, especially large ones, may not be as regular as the clathrate model suggests the general concept of more ordered, lower entropy water hydrating apolar molecules is widely accepted as the origin of the hydrophobic force.

Summary

The maximum entropy state specifies what will be observed at equilibrium, which is basically a state with no net forces acting at the macroscopic level. Entropic effects can give rise to forces. These forces appear when a system is not at equilibrium, or when part of a system is constrained from adopting its maximum entropy state, for example a compressed gas or a stretched rubber band. These forces are determined by the gradient in the spatial entropy contribution with respect to the pertinent displacement, e.g. a piston or the end of a polymer. Entropic forces are directed towards the maximum entropy state and close to equilibrium they are typically linear in the displacement. Entropic forces are distinguished from forces arising from interaction potentials by their linear temperature dependence. By introducing the entropic potential $-TS$, forces from either entropic effects or interaction potentials can be obtained in the same way as the negative gradient of a potential with respect to the relevant displacement. When entropic forces and interaction potential forces occur together, equilibrium may be described equivalently as

1. The state of maximum kinetic+spatial entropy
2. The point of balance between all forces, from entropy and interaction potentials
3. The minimum in free energy, where the free energy is a sum of a spatial entropy term multiplied by $-T$ and a potential energy or enthalpy term.

Chapter 5
Summary

Abstract Starting from the fact that atoms have both position and velocity, and using basic concepts of mechanics—force, kinetic energy, potential energy and conservation of energy—we have used complexion counting to obtain all the equations required to understand both entropy and elementary statistical mechanics, including free energy. These are summarized here along with their physical meaning.

Keywords Boltzmann factor · Partition function · Free energy · Equilibrium constant · Disorder

5.1 The Ubiquitous Boltzmann Factor

One mathematical factor appears over and over again in the discussion of entropy and statistical mechanics. It is the Boltzmann factor, which has the form

$$exp(X/k_bT) \tag{5.1}$$

where X is some quantity with the units of energy, scaled by the mean kinetic energy (or temperature). It first appeared in Eq. 1.15 describing the increase in number of velocity complexions upon adding heat. It next appeared in Eq. 1.18 describing thermal equilibrium between two objects in contact. It appears as Eq. 1.25 describing the probability of a particular spatial complexion. Most importantly, it appears as the weighting term for each spatial complexion in the expression for the partition function, Eq. 1.26. It appears as the factor describing the pressure distribution in the atmosphere (the Barometric equation 2.9). It appears in the Maxwell-Boltzmann distribution of atomic velocities, Eq. 2.14. Finally, it appears in the expression for the equilibrium constant, Eq. 3.19.

5.2   The Important Equations

The first important equation is the sum of the Boltzmann factors over all the spatial complexions, also known as the partition function, Eq. 1.26:

$$Z = \sum_{i=1}^{W_s} exp(-U_i/k_b T)$$

This quantity is central to statistical mechanics. This not surprising since, as Eq. 1.31 tells us, it is to within a multiplicative constant just the total number of complexions

$$W = C'Z$$

The Boltzmann factor itself, Eq. 1.25,

$$p_i = \frac{exp(-U_i/k_b T)}{Z}$$

gives us the probability of each spatial complexion. From this it is possible to calculate many average, or macroscopic, properties by forming their probability weighted sum over the spatial complexions.

   The entropy is, to within an additive constant, the logarithm of the total number of complexions. The entropy change given by Eq. 2.4 is thus the difference in the logarithm of the total number of complexions

$$\Delta S = k_b(ln W' - ln W)$$

For the special case of an ideal gas, the total number of complexions is just the product of the number of spatial complexions and kinetic energy complexions (Eq. 1.21). The former is proportional to the volume raised to the power of the number of atoms, V^N, while the latter is proportional to the total kinetic energy raised the power of $3N/2$, namely $E_K^{3N/2}$. This separability leads to a simple additive formula for the ideal gas entropy, Eq. 2.7

$$S_{id} = Nk_b ln V + \frac{3}{2} Nk_b ln T + C$$

The partition function can be used to obtain one of the most important average quantities, the mean potential energy, Eq. 1.28

$$U = \frac{1}{Z} \sum_{i=1}^{W_s} U_i exp(-U_i/k_b T)$$

The logarithm of the partition function (difference) yields a free energy (difference). For constant pressure conditions, this is the Gibb's free energy, Eq. 3.14:

$$\Delta G = -k_b T \, \Delta \ln Z = \Delta H - T \, \Delta S_s$$

where $\Delta H = \Delta U + P \Delta V$ is the change in heat (kinetic energy) resulting from the change in potential energy minus any pressure-volume work done. It should be remembered that the free energy change here is simply $-T$ times the total entropy change, which itself consists of a kinetic energy contribution, $\Delta S_v = -\Delta H / T$, and a spatial contribution, ΔS_s.

Finally, an important experimental quantity, the equilibrium constant, is also related to the free energy difference by Eq. 3.19:

$$K_{AB} = exp(-\Delta G_{AB} / k_b T)$$

5.3 What About Quantum Mechanics?

The treatment here uses classical mechanics only. Yet the real world obeys the rules of quantum mechanics. How does this change things? The answer is very little. First, Boltzmann's postulate still applies: All possible complexions (read quantum states) with a given number of atoms and total energy are equally likely (Tolman 1938). The definition of entropy as a logarithmic measure of the number of complexions still applies. Only the counting of complexions is affected by quantization. However, all the applications here involve large numbers of atoms at normal temperatures. Under these conditions, the number of quantum states is so vast and their energy levels so close together that the classical treatment of counting given here is more than adequate. The Boltzmann distribution is closely followed and the macroscopic, thermodynamic behavior is accurately described. Generally speaking, a quantum mechanical treatment is only necessary when studying a very small number of atoms, very low temperatures or the interaction of radiation with atoms.

5.4 Modest Entropies but Very Very Large Numbers

The machinery for complexion counting was applied to several everyday events. We now give both the change in number of complexions and the change in entropy, using Eq. 2.4

1. Heat flows from hot to cold. Temperature equilibration of two 0.1 kg blocks of copper differing in temperature by 1 millionth of a degree:

$$\frac{W'}{W} \approx 10^{3,400,000} \qquad \Delta S \approx 1 \times 10^{-16} \, \text{J/K}$$

2. A moving object comes to rest due to friction. For a 0.1 kg ball falling 0.1 m:

$$\frac{W'}{W} \approx 10^{10^{19}} \qquad \Delta S \approx 0.33 \, \text{mJ/K}$$

3. A gas expands into the available volume. For 10 cc of gas expanding to twice its volume:

$$\frac{W'}{W} \approx 10^{10^{20}} \qquad \Delta S \approx 2.6 \, \text{mJ/K}$$

4. A gas-cylinder spring of cross-sectional area 0.01 m^2 is loaded with a 4 kg weight

$$\frac{W'}{W} \approx 10^{1.4 \times 10^{19}} \qquad \Delta S \approx 0.43 \, \text{mJ/K}$$

5.5 Different Interpretations of Entropy

From its inception, entropy has been given a surprising variety of physical interpretations. The most common are discussed here in the context of the method of complexion counting described in this book.

5.5.1 Entropy as Disorder

Most of the pioneers in the development of statistical mechanics have at some time equated an increase in entropy with an increase in disorder. A completely non-technical but remarkably accurate description of entropy as disorder is provided by the anthropologist Gregory Bateson in a charming fictional dialogue with his daughter (Bateson 1972). Nevertheless, some care is required with this definition to avoid apparent contradictions. Take for example supercooled liquid water. Being a liquid, its molecular structure is quite disordered. Thermodynamic equilibrium is reached when the water freezes. Now the molecules are *more* ordered spatially as ice crystals. Yet this spontaneous process must have an overall increase in entropy. The contradiction is only apparent. When water freezes there is a positive latent heat of fusion which causes an increase in velocity complexions. Because the spatial component is literally the only part of the entropy change that is visible, it is easy to assume that it is the only part. This illustrates the limitation, alluded to in the introduction, of popular explanations of entropy that discuss only the spatial contribution, not the kinetic part. The latter, invisible, contribution must be included using Boltzmann's equation 1.15. The overall amount of disorder, spatial plus kinetic, does increase and the entropy change provides a quantitative measure of this.

5.5.2 Entropy as the Number of Microstates in a Macrostate

This definition is essentially that given in this book, if we equate a microstate with a complexion. I have, however, avoided the common term microstate, since it is subject to different usages, and it is often unclear whether it refers to just spatial complexions, velocity complexions or both. The macrostate, referred to here simply as the state of the system, is usually specified by a relatively small number of average quantities. These are typically observables such as volume, pressure, temperature/average energy, etc.

5.5.3 Entropy as a Loss of Information

In 1948 Shannon derived a measure of information H as

$$H = -\sum_i p_i ln p_i \tag{5.2}$$

The measure was derived using only general requirements such as consistency and additivity, yet remarkably it has the same form as the Gibbs-Einstein expression for spatial entropy, Eq. 3.10. For this reason Shannon called his measure information entropy. Jaynes (1957) saw this as more than a mathematical coincidence and asked how an information theory perspective could add to our understanding of thermodynamic entropy: Given knowledge of the macroscopic state of the system, what information does that provides us about the states of the system at the atomic or molecular level? The answer is that the thermodynamic entropy is a measure of this information. The extreme examples illustrate the point. If the entropy were zero, then knowledge of the macrostate would give us complete knowledge as there is just one corresponding complexion or microstate. Conversely, the maximum entropy state corresponds to the minimum amount of information: The number of alternative complexions the system might be in consistent with the observed state is maximal. A larger entropy means less information since we become less certain of the detailed state of the system.

5.5.4 Entropy as a Predictive Tool

A physical law is essentially a prediction that, given some appropriate conditions, something will happen a certain way. Although entropy arises from the random, statistical behavior of atoms, from the outset we have presented estimates of the vast number of complexions involved. The consequence is that the Second Law of Thermodynamics is truly law-like when anything more than a few thousand atoms are involved. Law-like in two aspects.

First, equilibrium behavior can with almost no error be predicted from the properties of the macrostate with the most complexions. Even if the system never samples that particular set of complexions it will with almost certainty sample *only* those indistinguishable from it. Here there is no requirement for ergodicity. Because of the vast numbers of complexions involved, the statistical limit to the accuracy of predictions about observable states and thermodynamic equilibria is better than that of most measuring instruments. The accuracy is in fact as good as, or better than, predictions of physical behaviors governed by non-statistical laws such as those of electromagnetism or classical dynamics.

Second, we can predict with almost certainty that a non-equilibrium system will change in such a way as to increase its entropy. In fact, provided the system is not too far from equilibrium the gradient of the entropy provides the forces, and hence the pathway of change. The famous tension between the time-reversibility of fundamental physical laws and the irreversibility of the Second Law was resolved, in Boltzmann's time, by recognizing the crucial role of initial conditions (Boltzmann 1895). Tracing the origins of lower entropy back further and further in time, the source of the time asymmetry that sets the direction for the Second Law is cosmological (Davies 1977). For reasons that are yet unclear, the universe started off in a very low entropy state because of the uniformity of the gravitational field, and it has been running uphill in entropy ever since.

5.5.5   Entropy as the Number of Ways Something Can Happen

The theme of this book is that entropy is simply a logarithmic measure of the number of ways the positions and velocities of the atoms can be arranged consistent with a given macroscopic, observable state. The corollary is: what we see is what can happen the most ways, what has the largest entropy. Even a very tiny entropy increase corresponds to truly huge increase in the number of complexions, ineluctably orienting the arrow of time we experience every day.

Appendix

A.1 The Exponential and Logarithm Functions

Aside from standard algebra, the only other mathematics used in this book involves the exponential and logarithm functions. Their basic properties are summarized here.

The exponential function, base-10, is written as $y = 10^x$, where x is the exponent. This is familiar from common notation used for large decimal numbers. For example

$$10^1 = 10, \quad 10^2 = 100, \quad 10^3 = 1000, \quad etc. \tag{A.1}$$

although the exponent need not be an integer. If two exponentials of the same base are multiplied/divided, their exponents are added/subtracted:

$$10^c \times 10^d = 10^{(c+d)}, \quad 10^c / 10^d = 10^{(c-d)} \tag{A.2}$$

The logarithm function is simply the inverse of the exponential function:

$$y = 10^x \Rightarrow x = log_{10}(y) \ or \ y = 10^{log_{10}(y)}. \tag{A.3}$$

Adding/subtracting logarithms multiplies/divides their arguments:

$$log_{10}(c) + log_{10}(d) = log_{10}(c \times d) \tag{A.4}$$

$$log_{10}(c) - log_{10}(d) = log_{10}(c/d) \tag{A.5}$$

Repeated application of the addition rule for the same number c gives the power rule

$$log_{10}(c^n) = n \, log_{10}(c) \tag{A.6}$$

© The Author(s), under exclusive license to Springer Nature Switzerland AG 2019
K. Sharp, *Entropy and the Tao of Counting*, SpringerBriefs in Physics,
https://doi.org/10.1007/978-3-030-35457-2

Base-10 is useful for expressing large numbers in a comprehensible manner: The exponent simply describes how many zeros follow the 1 in decimal notation. However the natural base for exponential functions (because of its more convenient mathematical properties) is Euler's number $e = 2.71828\ldots$. An exponential in this base is simply written as e^x. The corresponding logarithm to base e, known as the natural logarithm, is written as $ln(x)$. The rule for converting exponents between bases 10 and e is

$$e^x = 10^{x/ln(10)} \tag{A.7}$$

where $ln(10) = 2.303\ldots$. The corresponding rule for converting logarithms between bases 10 and e is

$$ln(x) = ln(10) \times log_{10}(x) \tag{A.8}$$

Exponents and logarithms in the natural base follow all the same rules as for base-10.

$$e^c \times e^d = e^{(c+d)} \tag{A.9}$$

$$y = e^x \Rightarrow x = ln(y) \quad or \quad y = e^{ln(y)} \tag{A.10}$$

$$ln(c) + ln(d) = ln(c \times d) \tag{A.11}$$

$$ln(c^n) = n \, ln(c) \tag{A.12}$$

A useful approximation for the natural logarithm when the argument is close to 1 is:

$$ln(1 + x) \approx x \quad (x \ll 1) \tag{A.13}$$

For example, when $x = 0.01$, $ln(1 + x) = 0.00995$ with less than half a percent error. For smaller values of x encountered here, the approximation is effectively error-less.

A.2   Mechanical Origins of the Ideal Gas Law

To round out the mechanical presentation of statistical mechanics the ideal gas law, also known as Boyle's law, is derived using basic mechanical concepts. This also provides the relationship between the mean kinetic energy of the atoms and temperature required in Sect. 1.2. Consider a rectangular box of volume V containing N molecules of an ideal gas. The molecules of the gas will continually be colliding with the walls of the box and rebounding. These collisions produce a

Fig. A.1 Origin of the ideal gas law

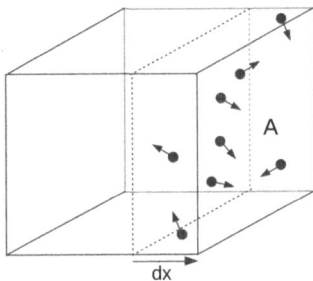

force on the walls given by the rate of change of momentum of the gas molecules. The pressure exerted by the gas is just the average force per unit area. Consider the molecules colliding with the right hand wall of the box (Fig. A.1). The area of the wall is A. If in a short time dt a number of molecules b with mass m and velocity u in the x direction collide with the wall, then the change in momentum is $2mub$. There's a factor of two because the collision reverses the velocity in the x direction. The pressure is the rate of change of momentum per unit area

$$P = \frac{2mub}{Adt} \tag{A.14}$$

The number of molecules b that collide with the wall in time dt is equal to the number of molecules within a distance $dx = udt$ of the wall that are moving *towards* the wall, namely with a positive velocity in the x-direction. This is half the total number of molecules in a volume Adx, since on average half the molecules will be moving to the right, and half to the left. The gas density is N/V, so $b = NAdx/2V$. Substituting for dx the number of collisions is

$$b = \frac{NAdtu}{2V} \tag{A.15}$$

Substituting for b in Eq. A.14, the pressure is

$$P = \frac{Nmu^2}{V} \tag{A.16}$$

Finally, since there will be a range of velocities, u^2 in Eq. A.16 must be replaced by the mean squared velocity, $\hat{u^2}$. This is obtained from the expression for the mean kinetic energy of the molecules, Eq. 1.4:

$$\hat{E}_K = \frac{m}{2N} \sum_i^N (u_i^2 + v_i^2 + w_i^2) = \frac{m}{2}(\hat{u^2} + \hat{v^2} + \hat{w^2}) \tag{A.17}$$

Now the mean squared velocities in the x, y and z directions will be the same since the gas is homogeneous and at equilibrium. So

$$\hat{u^2} = \hat{v^2} = \hat{w^2} = \frac{2\hat{E}_K}{3m}.$$ (A.18)

Using this expression for $\hat{u^2}$ in Eq. A.16 we obtain the ideal gas law

$$PV = N\frac{2}{3}\hat{E}_K$$ (A.19)

in terms of the mean kinetic energy. P, V and T had previously been measured for various gases leading to the empirical Boyle's Law

$$PV = nRT = Nk_bT$$ (A.20)

where n is the number of moles of gas and $R = 8.314$ J/K/mole is the gas constant. The second form of Boyle's Law uses the fact that the number of molecules in n moles is given by $N = N_a n$ where $N_a = 6.02 \times 10^{23}$ is Avogadro's number. So the scale factor relating temperature, measured on the ideal gas or Kelvin scale, and mean kinetic energy per molecule is

$$\frac{2}{3}\hat{E}_K = k_bT$$ (A.21)

where Boltzmann's constant $k_b = R/N_a$ has a value of 1.38×10^{-23} J/K.

References

Bateson G (1972) Metalogue: why do things get in a Muddle? In: Steps to an ecology of mind. Ballantine Books, New York

Ben-Naim A (1987) Solvation thermodynamics. Plenum Press, New York

Biot MA (1955) Variational principles in irreversible thermodynamics with application to viscoelasticity. Phys Rev 97:1463–1469

Boltzmann L (1877) Uber die Beziehung zwischen dem zweiten Hauptsatze der mechanischen Warmetheorie und der Wahrscheinlichkeitsrechnung respektive den Satzen uber das Warmgleichgewicht. Weiner Berichte 76:373–435

Boltzmann L (1895) On certain questions of the theory of gases. Nature 51:413–415

Bridgman PW (1972) The nature of thermodynamics. Harper and Brothers, New York

Craig NC (2005) Let's drive "driving force" out of chemistry. J Chem Educ 82:827

Davies PCW (1977) The physics of time asymmetry. Surrey University Press, London

Feynman RP (1972) Statistical mechanics; a set of lectures. W.A. Benjamin, Reading

Jaynes ET (1957) Information theory and statistical mechanics. Phys Rev 106:620–630

Shannon CE (1948) A mathematical theory of communication. Bell Syst Tech J 27:379–423

Sharp KA, Matschinsky F (2015) Translation of Ludwig Boltzmann's paper "On the Relationship between the Second Fundamental Theorem of the Mechanical Theory of Heat and Probability Calculations Regarding the Conditions for Thermal Equilibrium" (Wien. Ber. 1877, 76:373–435). Entropy 17:1971–2009

Tanford C (1961) Physical chemistry of macromolecules. Wiley, New York

Tanford C (1980) The hydrophobic effect, 2nd edn. Wiley, New York

Tolman RC (1938) The principles of statistical mechanics. Clarendon Press, Oxford

© The Author(s), under exclusive license to Springer Nature Switzerland AG 2019
K. Sharp, *Entropy and the Tao of Counting*, SpringerBriefs in Physics,
https://doi.org/10.1007/978-3-030-35457-2

Index

© The Author(s), under exclusive license to Springer Nature Switzerland AG 2019
K. Sharp, *Entropy and the Tao of Counting*, SpringerBriefs in Physics,
https://doi.org/10.1007/978-3-030-35457-2